The Metainterface

The Metainterface

The Art of Platforms, Cities, and Clouds

Christian Ulrik Andersen and Søren Bro Pold

The MIT Press
Cambridge, Massachusetts
London, England

This book was set in ITC Stone Serif Std by Toppan Best-set Premedia Limited.

Library of Congress Cataloging-in-Publication Data

Names: Andersen, Christian Ulrik, author. | Pold, Søren, author.
Title: The metainterface : the art of platforms, cities, and clouds / Christian Ulrik Andersen and Søren Bro Pold.
Description: Cambridge, MA : The MIT Press, [2018] | Includes bibliographical references and index.
Identifiers: LCCN 2017042793 | ISBN 9780262037945 (hardcover : alk. paper)
ISBN 9780262549677 (paperback)
Subjects: LCSH: User interfaces (Computer systems)--Philosophy. | Application software--Social aspects. | Human-computer interaction--Psychological aspects. | Computer art.
Classification: LCC QA76.9.U83 A524 2018 | DDC 005.4/37--dc23 LC record available at https://lccn.loc.gov/2017042793

Contents

Acknowledgments

This book is the result of years of research at the Department of Digital Design and Information Studies, School of Communication and Culture, Aarhus University. A special thanks to good colleagues and support from Aarhus Institute of Advanced Studies, Center for Literature Between Media, and Center for Participatory IT. We also owe great thanks to readers and reviewers who have provided invaluable suggestions and comments, including Geoff Cox and the anonymous peer reviewers. Also thanks to the great editorial team at MIT Press, Doug Sery, Katherine A. Almeida, and Noah J. Springer. Finally, thanks to our wives and families for sticking up with us during the long hours of writing and editing.

Fragments of the book are rewritten from published material in the anthologies *The Imaginary App*, *Critical Theory and Interaction Design,* and *Disrupting Business* and have been presented at conferences and festivals, including Electronic Literature Organization Conference, Transmediale festival for art and digital culture, Interface Politics 1st International Conference, Hybrid City Conference, and the Media Architecture Biennale.

Introduction

Art and Interfaces

To many people, computer interfaces relate to graphical user interfaces, and an implicit understanding of user friendliness and easy access to functionality. It is less known that since the early 1960s, there has been a tradition of artistic experimentation with computers, and this continuously informs a cultural and aesthetic critique—from the underbelly of digital culture, so to speak. These curatorial and artistic practices were first presented as part of an art scene that attempted to mix software with conceptual art. Exhibitions such as Cybernetic Serendipity at the Institute of Contemporary Arts in London in 1968, Software at the Jewish Museum in New York in 1970, and Computers and Visual Research in Zagreb in 1968 are all well-known examples of this. Within computer music, literature, games, and visuals, there are numerous illustrations of even earlier experimentation crossing computation with various forms of representation and performance, and today festivals such as transmediale in Berlin, Ars Electronica in Linz, and numerous other events attract tens of thousands of spectators and practitioners.

This book focuses on this art scene along with its relationship to interfaces, software, and computers, and more specifically the book's examples are commonly referred to as net art, software art, and electronic literature. In the words of Geert Lovink, such art forms "blow up the walls of the white cube ... in such a systematic manner that it moved itself outside of the art system altogether."[1] So although one may refer to the works presented in this book as art, art is in this case to be understood in its broadest sense. It may include what is recognized by the arts institutions and exhibited at museums, for instance, but it may also include what Gilbert Seldes in the 1920s referred to as the "lively arts"—meaning a scene that (like jazz, burlesque shows, or comics, as Seldes describes) finds its institutions outside high culture.[2] This by no means implies that the book rejects qualitative assertions about the works it presents, but only that the living practice is not necessarily performed by people who are trained at an art academy, and conversely that the artists in this field sometimes make works that are not necessarily recognized by arts institutions.

As indicated by concepts such as the postdigital, post-Internet, and new aesthetics, this breakdown between high art and a lively art of electronic networks and software also has to do with a normalization of these technologies.[3] The digital, online, networked, and so on, have become an intrinsic part of everyday life and routines, and it makes little sense to maintain that they have their own art form: all kinds of art—high or lively—now relate to and use digital technologies and platforms. What remains, as a cardinal point for this book, is the assertion that contemporary media and network technologies should be discussed as a cultural construct, and conversely that contemporary culture should be explored on technical terms. This is why this book concentrates on the relations between art and interfaces. The interface can be seen as an intrinsic and important part of the production of a contemporary condition where the capture and flow of data, information, and media is everywhere and part of everything. The interface as a technological and cultural construct is what brings about this contemporary world, and what makes it up to date and present at hand, real time, global, and so forth. In every sense, the development of tablets, fast networks, cloud computing, and more are not only technological advancements; they also seriously disturb and alter everyday cultural practices. The artistic practices that are examined in this book all relate to the production of this new reality.[4] But how does this take place?

Historically, artistic experimentation with computers and software has often been considered to be innovative. Though they do not always invent new technologies, artists may demonstrate the technologies' aesthetic and cultural value in new ways. For instance, in the 1990s with the advent of the World Wide Web, people experienced how text, music, and images were no longer protected cultural objects, but instead could be remixed and distributed freely in the network—rendering old technologies such as the landline telephone or music studio if not obsolete, then forced to undergo severe changes. Often one could see how the strategies applied in marketing or how the formation of new cultural products had been present in earlier examples of, say, net art. In this way, net art, software art, and other practices relate to what has been referred to as "disruptive innovation" and "creative destruction."[5] Whereas the mainframe computer and the Internet of the eighties were expensive and exclusive technologies, the PC and the WWW made them much cheaper and easy to access for ordinary people. This is common knowledge, but it is less known how artistic practices have sometimes led the way into new understandings and developments of this new consumer culture's expressions, preferences, productions, and other inner operations.

An example of this relationship between market innovation and artistic practice can be found in the works of net artist Christophe Bruno. In 2001, he launched the Internet installation *Fascinum* (figure 1) which displayed the ten most searched images on Yahoo portals in different countries in real time.[6] Already in 2004, Fabrica (Benetton's communication research center) had produced a similar installation (*10×10*), which ironically was later sold to Yahoo.[7] And in the French presidential campaign in 2007, the socialist candidate Ségolène Royal's website had a feature similar to *Fascinum*.[8] What is at stake for Bruno, however, is neither the alleged plagiarism nor the innovative potential of his art, but rather the ability of his practice to address its own subsumption. In his work *Artwar(e)* from 2010 (figure 2), he launches his own service that by means of a network analysis, creates artwork hype cycles as a form of "artistic risk management" and "computer-assisted" curating.[9] From the perspective of networks, *Fascinum*, *10×10*, and Ségolène Royal's website belong to the same hype cycle of "digital panopticism."[10] In other words, Bruno speculates in the particular networked mode of production, or inscription of his own work in a contemporary, global, and networked economy, which enables the transformation from "work" to "ware," and which he witnessed with *Fascinum*. By parodying and scrutinizing the language of the network, he further transforms this into a kind of "war" (as the title suggests).

The arts that deal with interfaces are, in other words, not just innovative. They do not belong in the realm of commercial products and services that people usually associate with computer interfaces, nor do museums or libraries institutionalize them. Rather, they are part of an arts scene that receives attention from both sides, and demonstrate an ability to reflect the larger conditions of a new regime of production. Borrowing a term from Walter Benjamin, one might say that they express a "tendency," as will be explained in chapter 1. Instead of using tendency as a reference to the banality of digital production, it can be associated with certain artistic practices' ability to show the networked computer's wider cultural implications on not only curating, marketing, and politics (as in the case of Bruno) but also reading, writing, working, knowing, and much more. Through their playfulness as well as interaction with interface enterprises such as Apple, Google, Facebook, or Amazon, they critically demonstrate how the networked computer imposes hypes, creates value, organizes labor, distributes cultural content, or inflicts new symbolic systems. Hence, this book seeks to bring this critique to the fore, and in its conclusion, reflect on how critique can be used in the design of interfaces, too.

In brief terms, this book is about interface aesthetics and culture, and as an analytic strategy, it focuses on the tendency in art that reflects the contemporary interface—that is, on readings of artworks such as Bruno's. In this sense, it presents contemporary

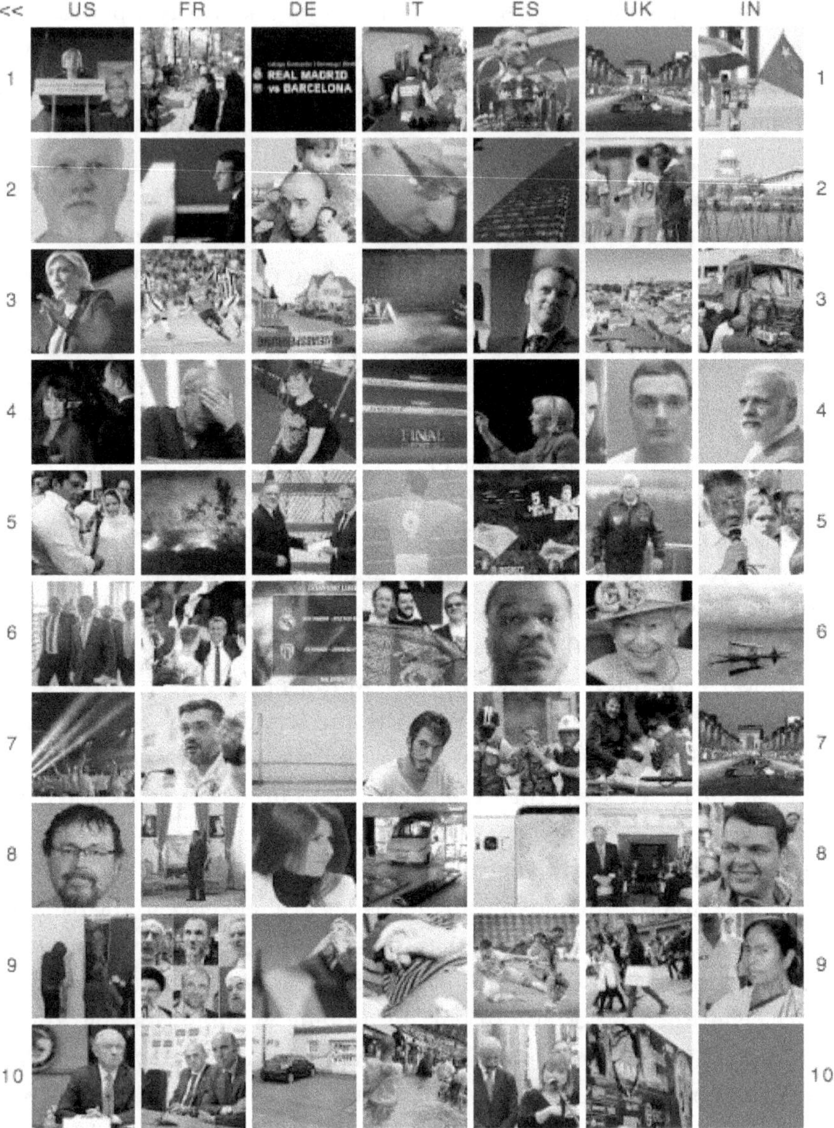

Figure 1

Fascinum (2001–) by Christophe Bruno. Real-time display of the most viewed news images (ranked from 1 to 10) on different national news feeds (Yahoo or Google). Courtesy of Christophe Bruno. Screen shot from web page.

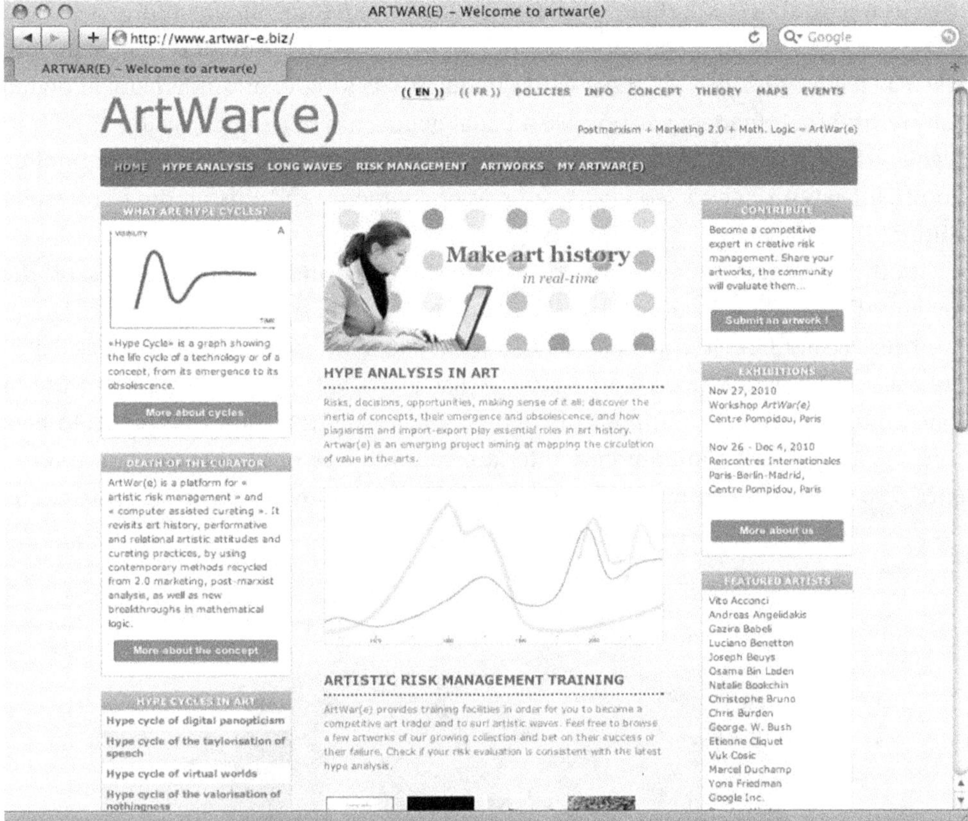

Figure 2

Artwar(e) (2012–) by Christophe Bruno. Web page that offers to calculate and map the circulation of value in the arts, and predict the hypes and trends of artworks and concepts. Courtesy of Christophe Bruno. Screen shot from web page.

artworks, but also reflects on the current challenges of contemporary interface culture in a situation where the computer's interface seemingly both becomes omnipresent and invisible, and where it at once is embedded in everyday objects and characterized by hidden exchanges of information between objects, or what it conceptualizes as a metainterface. The book, in other words, presents a new interface paradigm of cloud services, smartphones, data capture, and the like, and how particular art forms seek to reflect and explore this new paradigm. By bringing the tendency in artworks forward, the book aims to demonstrate how certain critical interfaces have an ability to reflect the deeper fissures within new technologies and the production of the work of art itself as well as show us an interface, after the interface has seemingly disappeared into "smart" futures and new promises of anticipation, participation, and emancipation. Like *Fascinum* and *Artwar(e)*, these works frequently belong to an underbelly of digital culture, and their tendency is not just a banal expression of a digital realm (as seen in Ségolène Royals website or *10×10*) but rather a critical exploration of a mode of production that feeds on networked circulations of phenomena, and this production's wider cultural consequences.

Another example of an artist who looks at the cultural consequences of the metainterface is César Escudero Andaluz. Andaluz is part of the Interface Culture LAB at Kunstuniversität Linz in Austria. His 2015 work *Inter_fight* (figure 3), which was developed at the Hangar.org visual art center in Barcelona, addresses the increased use of touch interfaces that like Bruno's distributed and "scale-free" networks, seem to be a core characteristic of contemporary interface culture's way of producing signification. *Inter_fight* consists of miniature robotic creatures that move around on tablets. The

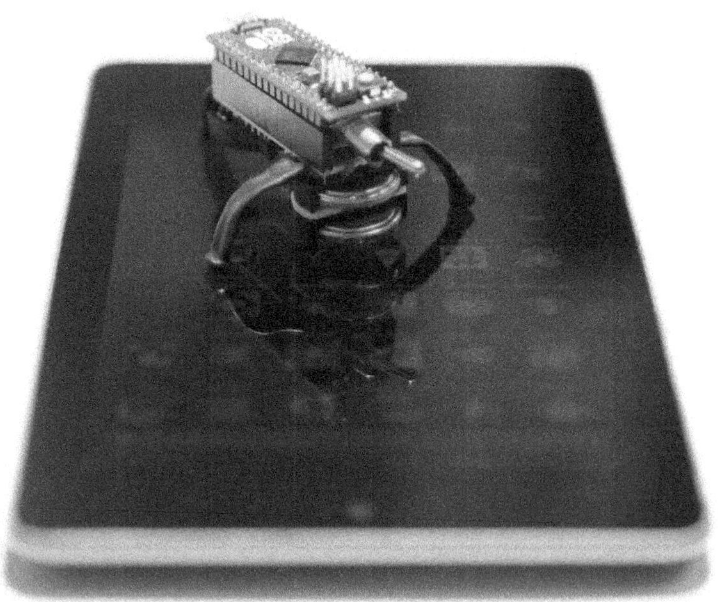

Figure 3
Inter_fight (2015–) by César Escudero Andaluz. Robotic creatures interacting with touch screens (sweeping, clicking, typing, etc.), and providing "false" information for website tracking. Courtesy of César Escudero Andaluz. Photo by César Escudero Andaluz.

creatures navigate by the light of the screen, and interact through conductive material in their "clicking" legs and "swiping" tails. The work is designed to "complicate the relationship between systems," and the direct interaction between the bots and tablets is in every way absurd. Their dance shows the functioning of the interface by disrupting the data capture of the websites they visit such as Google Maps. As Andaluz writes, "They provide wrong information for tracking website location, [and] fighting against the design homogenization and GUI standards."[11] In fact, they do so by letting one interface meet another. Besides conductive material, *Inter_fight* consists of vibrators, engines, photocells, and a microprocessor controlling the robotic creatures' behavior. In this way, they become interfaces that are added to the tablet's interface in a kind of closed cybernetic feedback loop. The apparent natural and invisible touch interface is revealed as a technical artifice that depends on signs and acquired gestures, and its capturing of behavioral data is demonstrated as a simple technique. Attaching an artistic interface to the interface and turning it into a closed cybernetic loop illustrates the artificiality of both the interaction and data capture that takes place on the Internet and in the metainterface.

Whereas both Bruno and Andaluz work on a formal-material level of the networked interface, and address production's inscription in networks on a coded or social level, the tendency can also be expressed in other ways, and address many different aspects of the material production of interfaces. As has also been pointed out by Matthew G. Kirschenbaum, interfaces have a "forensic" materiality that refers to how data are materially inscribed in computers in ways that cannot be deleted.[12] These layers can be brought to the fore, or excavated, in forensic readings of electronic literature (as Kirschenbaum does), and point to the interface's material topology (hard drives, cables, and so forth). As an alternative to practices that highlight the histories of media technologies, and practices that turn the material architecture into a formality (that carries content in new ways, as digital or new media), these archaeological and forensic practices point to the traces and fractures that media technologies leave behind.[13] They reflect the legacies of computing systems (file formats, protocols, etc.) as remains or ruined monuments that may, for instance, bear witness to preferences, values, ideologies, (geo)politics, and so on. Contemporary metainterfaces also have legacies or a "prehistory," as, for example, Tung-Hui Hu argues in relation to cloud computing. This prehistory of the cloud points to not only the legacy of former infrastructures and spatial evidences of state or corporate power (e.g., how Cold War bunkers were turned into data centers) but also how this history is displaced.[14] As much as the cloud is a new service in contemporary computing, it is also an interface construct that produces a distance to the material architectures of the network, a particular "clouded" perception of the contemporary world. This construct is a central characteristic of the metainterface.

The Barcelona-based artist Joana Moll's project *CO2GLE* from 2015 is a good example of a work that points to the material evidences of interfaces (figure 4). It specifically alludes to the geologic age of the Anthropocene, and how the use of interfaces, as a human activity, influences the climate and environment. One second after opening, the web page *CO2GLE* simply states that "Google.com emitted 510.49 kg of CO2 since you opened this page," with simple black text on a white background, and it keeps adding 510.49 kilograms every second.[15] Google.com is the most visited website on the Internet, and according to Moll, the Internet is responsible for 2 percent of the global carbon dioxide emissions (which by comparison is equivalent to aviation).[16] In other words, though the perception of the online as a virtual cyberreality is persistent, and often people tend to think of interfaces to web services and the kinds of activities they

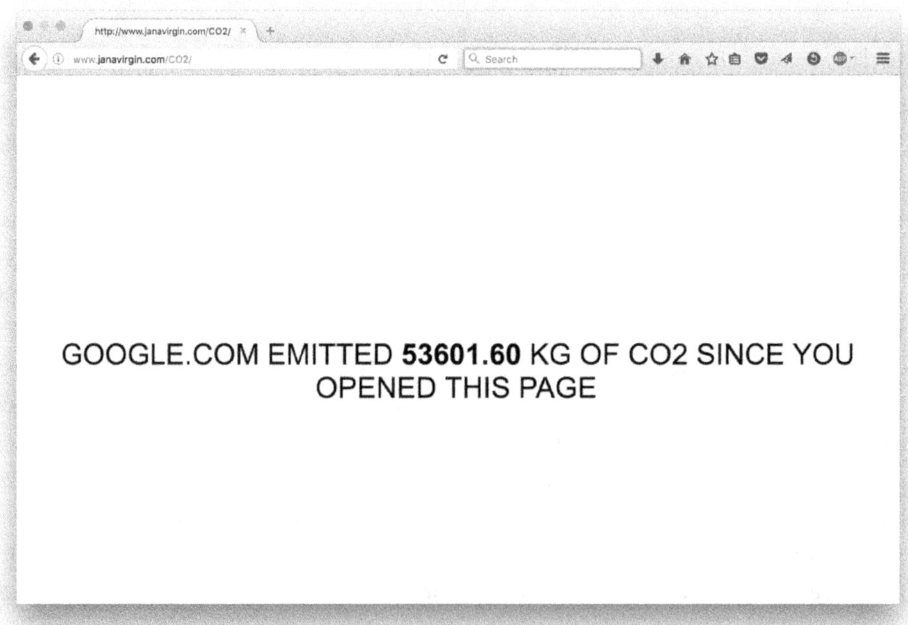

Figure 4
CO2GLE (2015–) by Joana Moll. A real-time, net-based installation that displays the amount of CO_2 emitted each second by Google.com. Courtesy of Joana Moll. Screen shot from project website.

perform with them as immaterial, they clearly have material consequences: when people search on Google, or use other networked services like Dropbox or Facebook, they pollute. As the actual emission might happen far away in a different country, people do not necessarily sense the effects in their immediate environment. This, however, only points to a shift of scale that might be difficult to comprehend, but that is no less real. To put it another way, the material workings of the Internet are often displaced, but the work foregrounds the networked interface's infrastructure and global implications on the climate. Even though these climatic consequences of the networked interface happen somewhere else and, so to speak, are concealed by the interface, *CO2GLE* demonstrates that they are still real and significant.

Parody, intervention, and excavation, along with many other artistic modes of exploration, function as ways to reflect the work's material levels and production, and through this, also the wider cultural consequences of the interface's new regime of production. Bruno, Andaluz, and Moll each in their own way address contemporary interface technologies and aesthetics, which are at once symbolic and material as well as undergoing severe changes. Finally, this book is not only about the art of contemporary interface culture that reflects such cultural changes. It is also a design book that actively seeks to speculate on the potential of applying a critical perspective in interface design, and how to actively relate to and reflect the fissures of production in interface design.

The Metainterface

In the 1990s and early 2000s, when the interface to many was still considered a work-related phenomenon (a legacy of the 1980s' introduction of interfaces into the workplace), net and software art enabled understandings of how desktops, web browsers, image editors, and so forth, were also cultural constructs. In other words, net and software art showed how the interface did not just disappear behind office and work metaphors along with desires of seamless interaction but resurfaced as a new milieu for cultural practice, too. Many artworks pointed to how values and conduct were embedded in technical infrastructures, and how PCs and the World Wide Web had become not only new production and dissemination systems for existing cultural objects (text, images, movies, etc.) but also the conveyor of new cultural practices that included sharing, remix, and much more. *Fascinum*, and the ways it presents visual culture, is a good illustration of this. Other works deconstructed the usability of image editors (such as Adrian Ward's *Auto-Illustrator* [2000–2002]), or pointed to new commons-based cultural production that challenged existing notions of origin and copyright (such as Cornelia Sollfrank's *Net.art Generator* [1997–]).

In many ways, the interfaces of today are much more embedded in the kinds of cultural practices that were highlighted in net and software art. To put it simply, though the desktop and PC still exist, the interface has moved from the office into culture. The new platforms for this culture are the ever-present media devices and apps, and displaced networks of clouds and data streams. It is this shift, and how the interface disappears, not into seamless work-related activities, but into the environment and everyday cultural practices, that is the topic of this book, conceptualized as a metainterface. Whereas the work-related interface was designed to disappear as windows into virtual and immaterial realities and workspaces, today's cultural interfaces disappear by blending immaculately into the environment via what is commonly referred to as mobile computing, ubiquitous computing, and the Internet of things, among other terms. The same way seamless immersion is the cultural promise of the screen-based interface technologies, social promises such as the networked, participatory, or open (which were once solemnly associated with the underbelly of net culture) have now become intrinsic to the development of interface technologies. The promise to deliver interaction spaces where apps, smart objects, and mobile computers blend flawlessly into the environment is a promise to deliver a smart, networked, participatory, and open future.

The central claim of this book, however, is that despite the attempts to make the interface disappear, and conceal it behind a mask of smoothness and real-time information flows that all seem to be for the social, individual, or functional good, it gradually resurfaces.[17] Although the interface may seem to evade perception, and become global (everywhere) and generalized (in everything), it still holds a textuality: there still is a metainterface to the displaced interface. Artistic practices may help us to see this by reflecting the fissures in the kinds of realities that the metainterface produces. That is, they may reflect their own production, and in this way, challenge the disappearance of the interface, and make it much less innocent than corporate rhetoric otherwise suggests: the ubiquity, pervasiveness, and real-time smoothness of the interface are loaded with worldviews, values, ideologies, politics, regulation, and conflicts, and thus also hold within them potential new forms of expressiveness. The concept of metainterface consequently performs three different, but related, functions.

First, it is the description of a contemporary interface paradigm. One may argue that the layered structure of the computer (from assembly language to graphical user interface) makes any interface a metainterface, but what is specific about the current interface paradigm is the universally dispersed, omnipresent nature of this. As such, it is an interface to the many hidden interfaces and clouded exchanges of data and signals in a series of platforms that connects the everyday use of apps on a smartphone to large-scale, globally networked infrastructures.

Second, the industry around the metainterface presents this as a new reality of smooth access and smart interaction. Yet its protocols also hold within them reconfigurations of the everyday production, distribution, and consumption of culture as well as the reorganizations and renegotiations of urban space, and profound changes in the relation to material infrastructures that also affect sense perception itself—that is, the perception of a globalized and real-time world.

Third, the notion therefore points to metainterfacial artworks along with ways of analyzing them as critical explorations of their own material conditions that not only reflect the metainterface industry's corporate production of new realities but also depicts alternative ways of constructing and designing the metainterface. One may, for instance, experience how the interface resurfaces as absurd predictions, as in the case in *Artwar(e)*; the surreal use of maps and exchanges of data, as in *Inter_fight*; or displaced material infrastructures as in *CO2GLE*. The works, each in their own way, point to the reality that metainterfaces produce, and how they reconfigure a range of domains from cultural production to urban space as well as the perception of a globalized world. Such explorations may potentially lead the way to the design of new interface paradigms, too.

In other words, the artworks at the core of this book reflect the fissures and crises of production of an interface that moves beyond the networked PC, and they each highlight three different kinds of material symbolism of this new interface. Besides a theoretical introduction, and a concluding reflection on how to apply this in critical interface design, the book's chapters address these three material tendencies, including a semantic capitalization of language (as seen in the apps and devices of a new interface-based culture industry, and illustrated by Bruno's focus on the networked character of cultural production and consumption), semantic territorial control (as seen in urban interfaces, and shown by Andaluz's emphasis on the inscription of space), and a virtualization of the material architectures of the network that has deeper effects on the experience of reality (as seen in the cloud interface, and characterized by Moll's focus on the Anthropocene). Though these tendencies seem dispersed (and the chapters are also readable independent of each other), this book argues that they are related at the level of interface aesthetics: they are related at the level of the language and grammar of a new material textuality. The chapters will progressively explain how this textuality functions as a formal inscription of reading and writing (in cultural devices and platforms) that unfolds as a spatial and territorial inscription (in cities) that assumes immense and incomprehensible material scale with consequences for sense perception (in the cloud). The final chapter will discuss how this critical understanding of the metainterface's material textuality is not only a way to discuss art and interfaces but also a way to conceptualize and design, beyond a simple belief in the disappearance of the interface.

The first chapter describes the theoretical framework of the book. This includes an elaboration of the book's key notions, notably the interface itself, and an explanation of why it is an important beginning for an aesthetic critique; an elaboration of "tendency" as a key concept that address material production and textuality, and enables new understandings of artistic practice with interfaces; and finally, an analysis of how the interface is changing and becoming a metainterface. From being a relatively self-contained entity, it is now mobile, ubiquitous, and deeply entangled with massive data capture and techniques that make the interface appear intelligent. Artistic practices can be regarded as an occupation that reflects and reflects on this tendency, and more specifically, how the interface reappears as a metainterface, with a material textuality and particular grammar that reflects linguistic, territorial, and experiential crises in production. The following chapters will analyze these changed material relations and crises in the production of the artwork, and relate them to a wider cultural critique of interface culture.

The second chapter addresses the new platforms of a cultural metainterface industry (the smartphone, e-reader, etc.). These platforms, which are central to the metainterface, are built on a capitalization of net culture. As much as this opens up new possibilities for cultural production, it is also experienced as a crisis of black boxing and tracking. More specifically, the metainterface industry is based on a "semio-capitalism" that involves a formal inscription of not only production (e-books, streaming of movies and music, etc.) on the platforms but also consumption. What people look at, read, listen to, and so on, becomes data that can be used to anticipate user behavior. In other words, the loss of access to the media technologies that characterizes the metainterface is a loss of ownership and privacy. This is of political importance, but it also relates to what writing and reading is and becomes, as a partly nonhuman activity. The metainterface industry transforms reading and writing: consumption, which traditionally has been considered passive, becomes a productive inscription of behavioral data, and the production of culture becomes a kind of consumption. The interface itself is no longer just a consumer product that can be bought or sold in a traditional sense, but is a networked streaming service that is financed by the capturing and inscription of user behavior.

The third chapter examines the urban metainterface. Apart from houses, factories, offices, streets, electricity networks, sewage, and so on, a city can be defined as a "semiotization" of space. Media and language—including street signs, advertisements, and more—also inscribe space. In other words, language is intrinsically related to an ongoing process of territorialization in the urban. Naturally, changes in the linguistic apparatus, as those appearing with the metainterface and its language-based industry (also known as semio-capitalism), will also affect this process. The chapter thus explores how the metainterface industry disrupts the everyday practices of urban life, and how the city is read, perceived, inhabited, and organized in new ways. Immediate examples of this are phenomena like Airbnb, Uber, and TripAdvisor, which all point to how anything potentially can signify anything: an apartment can be a hotel, a passing car can be a taxi, and so forth. To read and use the city, one only needs the right app. The chapter will discuss how the city is opened up by the metainterface industry, but also how this implies a new organization of the city, and how the mechanisms of semiotization (which determines how the city should be read, interpreted, and used) become increasingly opaque as well as layered with an algorithmic grammar and way of seeing. The process of opening up the urban sphere with the interface hence also implies a new mode of control where alternatives to "the open" are needed.

The fourth chapter addresses the clouds of the contemporary metainterface, and how cloud computing influences the experience of a world where the interface is in everything (generalized) and everywhere (global). In other words, how the interface becomes a metainterface, or how the interface seemingly disappears but resurfaces as a metainterface with its own textuality and grammar. If the desktop was the metaphor of interfaces in the 1990s, the cloud is emblematic of the cultural interfaces of today.[18] Cloud computing typically covers services that offer data storage (including e-mails, photo libraries, text documents, and much more), and also the power to process data, which are left "out there," "in the cloud." What were once network architectures and material inscriptions that the users who built the World Wide Web cared for as well as cherished, have now become standardized and displaced out of human reach. The main objective of this chapter is to explore how this material displacement influences the experience of the interface as an alienating smoothness and phantasmagoria, and how this experience of the *meta*interface affects sense perception at a more general level, too. With the war on terror and climate change as examples, the chapter thus examines how the metainterface becomes emblematic of the experience, knowledge, and actions of a global, synchronous, and real-time world.

The final chapter concludes by discussing how to extend an interface criticism into a new tradition of design, or how to design with an awareness of the language and grammar of interfaces. Although the metainterface is intrinsic to new cultural industries, a new de- and reterritorialization of the urban, and new experiences of realities, there is no reason to take it for granted. As the examples in this book hopefully will demonstrate, seeing and writing with the metainterface can allow for experiences of surrealisms, beauty, anxieties, laughter, and much more. In a contemporary technology critique, it is well worth knowing how the smoothness, phantasmagorias, and prophecies of openness and participation function as interface constructions that combine semantic processes of signification with machinic processes of signals. Nevertheless, giving way for such freedom is not only an analytic insight; it can also fuel new technological imaginaries, and may function as a catalyst for a design paradigm. In other words, the subtitle of this book (*The Art of Platforms, Cities, and Clouds*) points not only to how computing today makes way for new artistic expressions but also to how computing and interface design itself may be practiced, "tactically" as "not-just-art" (as Matthew Fuller once characterized software art).[19] The final chapter thus presents two case studies and discusses this interface criticism by design in relation to a larger history of critical design, particularly the Scandinavian political tradition of participatory codesign and Anglo-Saxon tradition of critical design.

The selection of the cultural metainterface industry, urban interface, cloud interface, and critical design of interfaces as the main foci in the book by no means expresses an intention to exhaust an interface criticism. Naturally, many other areas could have been included as they too are affected by the metainterface (e.g., the body, logistical infrastructures, networked activism, etc.). By presenting a perspective, analytic strategy, and practical tactics, however, this book will hopefully open up the field of interface aesthetics and criticism, rather than restrain it to certain phenomena. As much as it seeks to cover important domains and tendencies that interest its authors, it hopefully outlines a methodology for a contemporary cultural critique and design of interface culture in a time where interfaces are ever more present, yet ever more unintelligible.

1 Interface Criticism: Why a Theory of the Interface?

Any understanding of the metainterface's cultural and aesthetic significance must begin with a clarification of what an interface is, and why it is an important notion: why the interface is inevitable for the networked computer, and why there always will be an interface—even in a situation where it has become displaced (seamless), global (everywhere), and generalized (in everything). Interfaces and their aesthetics have been the object of our studies for more than a decade. We, as authors of this book, and like many others in the late 1990s and early 2000s, initially set out with the intent to research the relationship between literature and computers. With its algorithmic text processing, and complex relations between code and text, the computer in many ways seemed to offer new perspectives on, for instance, textuality, and also instigated new forms of text, such as electronic literature and computer games with epic narrative elements. Neither electronic literature nor computer games compare to printed books, however, and conventional literary textual analysis does not apply easily. There was a new materiality to the textual that called for new theoretical approaches.

This discrepancy between textual organization and its analytic apparatus was also coined by Espen Aarseth in his groundbreaking work *Cybertext: Perspectives on Ergodic Literature* from 1997.[1] According to Aarseth, the new formats for electronic literature that were found in hypertext fictions or Multi-User Dungeons were not manifestations of a postmodern notion of the text as a rhizome, as much poststructuralist hypertext theory argued.[2] For instance, the user of a cybertext was a "more integrated figure than even reader-response theorists would claim."[3] The kinds of interaction that cybertexts offered were different from the traditional conventions of literature and the printed book. Whereas the flipping of pages in a book is trivial next to the interpretative interaction when reading a book, it can almost be the other way around when interacting with hypertext and cybertext. In cybertexts, the interaction became "ergodic" as a way of working out one's path through the text, and the cybertext literally becomes a "machine for the production of variety of expression."[4] As the term "cyber" indicates (originating in the Greek *Kybernetes*, which means steersman), navigation is in many cases crucial. For example, to many computer gamers the narrative is only interesting insofar as it explains what to do next. As Aarseth pointed out, such text forms do have a history and tradition, as seen, say, in the writings of the French literary movement Oulipo (Ouvroir de littérature potentielle, or workshop of potential literature), which used conceptual or material constraints in its writing and produced cybertexts such as Raymond Queneau's *Cent mille milliards de poèmes* (Hundred thousand billion poems), but this understanding of text and its history had usually been marginalized. Both electronic literature and computer games were, in other words, not just new media for narratives but also presented a type of text that was characterized by a schism between doing and reading, or instrument and media. These texts offered literary expressions that were constantly interrupted by the presence of ergodic action. This influence of the organization of the text had rarely been an issue in literary analysis and textual theory, but Aarseth made it clear why cybertexts and ergodic literature called for new theories.

Rather than inventing new names for this particular type of text and its artistic genres, one may choose to stick to the colloquial term interface that most people are familiar with, and refer to the genres and art forms by the names that have most often been used, such as computer games, electronic literature, net art, and software art. All these arts are interface based, and frequently also follow parallel historical developments of exploration of the nonlinear or computer-generated expression. They take advantage of multimedia, too, and blend into one another to the point where it becomes difficult and sometimes useless to sustain the borders between the fields of literature, art, and music: images include texts, poetry is performed as readings with music, and so forth. In addition to existing aesthetic theories, there was, as Aarseth noted, and still is, a need for conceptual and material understandings of these artistic interfaces along with their formal expressions, histories, and interplay with broader cultural characteristics. This book is a contribution to this field, and provides an insight into the arts of interfaces, and how they relate to contemporary developments in an information technology that by all means is a technology that both builds on and transgresses contemporary aesthetics and culture.

This chapter sets the theoretical frame of this book and delves into what interfaces are as cultural-technological constructs. It also explains how they can be critiqued, not just for being good or bad, useful or useless, right or wrong, beautiful or ugly, but for their own literary value (as Aarseth recommends). Beyond the moral of what they convey, their ease of use, and the logic of their operations, they have an aesthetics that calls for a new literary interface critique, too: through reflection of their own means of production, they have the potential to reflect how interfaces and networked computers change culture in a larger perspective, and the purpose of this book is to discuss these aspects of the interface and interface culture. In other words, this chapter explains why and how artistic practices express this tendency. Furthermore, it includes a reflection on how the interface is changing from being relatively contained in software and hardware, to being mobile, limitless, and almost reckless and dissolute in the ways it incorporates a constant flow of data from other interfaces. This change fundamentally presents a new mode of production that is, as will be presented in the following chapters, setting the stage for the tendency in contemporary art that deals with interfaces as well as networked computers and platforms.

Theories of the Interface

A theory of the interface is of course not a new thing, and the textual assemblage of representation and organization that Aarseth also described has received attention in several different ways throughout the past decades. In their seminal book *Remediation* from 1999, Jay David Bolter and Richard Grusin depicted the assemblage as a perceptual friction between the logics of "hypermediacy" and "transparency," or "looking at" and "looking through," which have origins stretching back to the Renaissance and the invention of a linear perspective in Western visual culture.[5] A few years later, in 2001, Lev Manovich in *The Language of New Media* explained how the language of cultural interfaces draws on traditions of media such as movies and written texts as well as instruments and interfaces to machines.[6] In a wider perspective, all software interfaces are dealing with the kinds of composition that Aarseth, Bolter and Grusin, and Manovich have described in their own ways: software interfaces generally convey meaning and action, representation and computation, media and instrument, and so forth. An icon on a desktop is, for instance, at once an image and an access to an operation in the computer.

The various theories of interfaces from the late 1990s and first decade of the new millennium all reflect how media theory adopts concepts from computer science (such as the interface) as central categories, and how media theory becomes software studies. This development was coined in a US tradition by Manovich, and in Europe, predominantly by an arts scene. The most prominent examples of this scene were perhaps witnessed at the Readme software art festivals organized by Alexei Shulgin and Olga Goriunova together with a wider international group in Moscow (2002), Helsinki (2003), Aarhus (2004), and Dortmund (2005), but were also an important part of the Transmediale festival in Berlin under the leadership of Andreas Broeckmann

(2001–2007). Several important texts and books were published at the time, such as the 2008 *Software Studies: A Lexicon*, which was edited by Matthew Fuller.[7] This tradition also laid the foundation for our attempt to coin an interface criticism in our 2011 book, *Interface Criticism: Aesthetics beyond Buttons*.[8]

In a time of smartphones, smart cities, and many other phenomena, the interface has seemingly become ubiquitous and networked in the sense that it is increasingly everywhere and limitlessly connected. This outlines the continued necessity of developing an understanding of the interface, and recently there have been several efforts to do this. For instance, there is Alexander R. Galloway's political perspective on the processes of interfaces, Lori Emerson's literary perspective on how writers address the interfaces of reading and writing from the book and typewriter to the tablet, and Branden Hookway's philosophical perspective on the interface as a relationship with technology.[9] The interface is also conceptualized in different ways—notably as "stacks" by Benjamin Bratton, signifying the layered structure of an interface that is at once technical, cultural, and part of a new political reality.[10] In addition to such efforts, this book focuses on the interface as a material and technical format that juxtaposes the operational with the representational, and thus is deeply entangled with the cultural and aesthetic domain. Wendy Chun's account of metaphors, direct manipulation, ideologies, and real-time adaptation in interfaces also seems particularly useful in this context.[11] This juxtaposition of instrument and media, and how it is played out in artistic practice, is core to the understanding of interfaces that are applied in this book. But it also expands the notion of interfaces into an understanding of current developments that are exemplified by the phenomena of platforms such as the e-reader and smartphone—interfaces to the urban and cloud computing, where interfaces become networked in ways that are sometimes hard to comprehend.

This book goes beyond its literary starting point, and includes the different art forms and studies of software that relate to the interface; still, it is the literary perspective that guides its interests. Though text is fundamentally changing when it becomes computational, there are still textual structures such as inscriptions, grammars, scripts, and codes, and it is therefore relevant to draw on not only the literary theories of new textual understandings (such as Aarseth, N. Katherine Hayles, and Manuel Portela) but also the more classical materialist literary theories of Walter Benjamin and others.[12] Like these theoreticians, we, as authors, do not confine ourselves to the field of literature but instead see literature as part of a broader field of artistic and material production that includes artistic practices with networked computers. In a theoretical discussion of digital media text forms, the challenge is hence to transgress the limited understanding of text as a formal system of representation, and pay attention to the particular formal, contextual, and technical materiality of the algorithmic text. In this sense, the theorization of the interface presented in this book differs from poststructuralist semiology and semiotics' conception of a generalized textuality of signs without materiality.[13] Instead, interfaces are discussed as a new form of textuality with other grammars and inscriptions than traditional text. As the practice-oriented research professor for new communication technologies Florian Cramer has pointed out, computers can be seen as textual machines whose material properties and contexts differ from other texts and textual machines in that they run on "alphabetic code."[14] Following on from Cramer, it should be discussed how the interface has changed the material, concept, and role of text, rather than whether interfaces fit within existing notions of text and literature. The interface thus becomes a text in an analytic strategy, but a text that is material, technological, and part of more extensive political and social contexts than what existing notions of text traditionally point at.

Without sacrificing notions of representation and textuality, the understanding of text applied in this book is fundamentally to be understood in its relation to the material processes of the machine. The approach is related to current theoretical interests in materialism such as new materialism as well as post- and nonhuman theories, but this book insists on stressing how representational and material dimensions are intertwined in interfaces. A critical interface theory relies on the understanding of discrete ideological and political mechanisms that are often hidden within the implicit representations and grammars of the interface. As argued by David Berry, and also others, contemporary materialism such as speculative realism or object-oriented ontology frequently tend to disregard such dimensions.[15] In contrast, the materialism that this book seeks to assert is inherently dialectical, and explores the relations between material and mediation, including the ideological, political, and cultural dimensions of this relation.[16]

Critical analyses of technology, including analyzing technology's cultural, ideological, and social dimensions, are important parts of this book's approach. Accordingly, it shares many interests with science and technology studies. Although the French sociologist and philosopher Bruno Latour has been criticized for pluralizing materialism to a level of abstraction, his discussions of how phenomena become "Things" or "matters of concern," as presented in his collaboration with the German artist and curator Peter Weibel at ZKM / Zentrum für Kunst und Medien (Center for Art and Media) in Karlsruhe, are good examples of this.[17] The book is without question also related to Donna Haraway's political, feminist voice in science and technology studies, especially her suggestion of a mediated perspective that includes the technological body as a "material-semiotic actor" and avoids reducing technology to a mere material object of observation.[18] Haraway's material-semiotic actor relates to the observer (as a techno-material cyborg), but also to the media, instruments, and technologies through which we observe, and in this sense it corresponds closely with interface criticism, which like Haraway's feminist objectivity is a material and situated knowledge.

Taking as its starting point a literary approach to the material semiotics of interfaces, the theoretical outline necessarily begins with theorizing the networked computer's fundamental ability to combine processes of signification with the signal processes. Although the book's literary materialist approach also differs from computer semiotics' understanding of interfaces, there is still something to be gained from the field and its understanding of how the interface combines representational and instrumental dimensions.

Computer Semiotics and Interface Design

In the theories of human–computer interaction (HCI), the juxtaposing of instrument and media has historically been considered a fundamental challenge for interface design. To overcome the challenge, interface design has often aimed at concealing not only the inner processes of interfaces (e.g., the running code and its command lines) that makes the interfaces inaccessible to ordinary users but also its hidden processes of labor.

Computers are not ordinary machines; they are symbolic ones, as computer semiotician Peter Bøgh Andersen claimed in the late 1980s: they are constructed and controlled by means of signs, or expressed differently, they are constructed by interfaces. The graphical user interface is an obvious case of how machine signals are controlled by means of signs, but underneath the graphical interface there are other interfaces. There is, for instance, a program text that contains signs that stand for possible program executions. The executions also involve a compiler, or interpreter. A compiler is a program that transforms source code into computer language, or object code, that can run on the computers central processing unit. In other words, it enables the execution of the source code by means of text. As Andersen explains, "Everything in a computer system, from top to bottom, is used as signs by some group of professionals. At each level, there are texts that must be interpreted a statements or prescriptions about some present or future state of the system."[19] To put it briefly, a basic assumption in this book is that there is no computer without an interface. The interface can be the well-known graphical user interface between user and computer, but it exists at other levels, too: between different programs, and even between hardware components. Yet there is no privileged interface either: at all levels, the interface combines signs with signals, and at all levels it is saturated by choices, conduct, language, values, and worldviews, and consequently, also an aesthetics that reflect these saturations critically.

Though a symbolic machine, however, the computer processes signs in a far different way than a human interpreter. Frieder Nake, who is not only known for being a pioneer in computer art but also a computer semiotician, has explained the difference like this: "Whenever we program a computer, what we are really doing is mechanizing mental labor. In order to mechanize a specific mental labor, it must be generalized in the broadest sense possible, it must be formalized, and finally turned into computable form. ... The mental labor remains with the human as their genuine capacity. ... But besides this human capacity, we encounter its algorithmic form."[20] This distinction between mental labor and its formalization is crucial. All programing of software is always an expression of a process that reflects a formalized relation to labor—that is, a translation of mental labor into a computer program. I, the human being, can interpret the world as signs. The computer, on the other hand, is stupid and only capable of determining, not understanding. Everything must be computable, and hence the computer can only interpret things at this basic level.

Following this line of thinking, the computer is conceptual by nature. Any computer program is a manifestation of a concept: I "think" searching, sorting, listing, and so on, and I make it into a machine that can do it indefinitely. Labor therefore has both a mental form (meaning a tacit knowledge of the process of, for instance, searching, sorting, or creating) *and* an algorithmic form (searching, sorting's symbolic representation, expressed in formal instructions of, say, sequencing, iterating, and selecting, as they are often applied in programming).

At the level of the graphical user interface, the formalization of labor gets doubled and is represented to the user as signs that can be used to manipulate mental labor. For example, an interface for searching a file in an archive is based on both the formalization of the mental labor process of searching in the form of algorithms and a representation of this to the user in the form of a text field along with an icon displaying a looking glass.

This sort of interface design (understood as the successful mapping of the formalized labor process to the user's world of signs) will in many cases cause a forgetting of the interpretation of reality involved in the programmers' formalization of the world as well as the computing and signal processing that takes place within the computer when executing a search. Even the simplest act "sets into motion huge piles of frozen mental labor that others—system and software designers and programmers have done. ... The user can hardly escape the impression it is the computer acting all by itself, in particular producing signs," Nake writes.[21]

In other words, when users search using Google or other search interfaces, they activate labor processes. Rather than systematically looking through and organizing the archive, they (in order to save time) rely on a predefined formalization, and a machine that can interpret this language and determine an outcome. In this process, they often tend to forget the conceptual and performative elements that are involved in actions, and that do not strictly speaking make it the user's search; it is rather the software and its technical infrastructures that search. Google's search algorithms are in fact also renowned for their obscurity and continual updates, and likewise most people know little of their servers and the network that brings the results to them.

Interface Aesthetics

Following on from computer semiotics and the field of HCI, interface aesthetics is frequently reduced to a pragmatic purpose, or an unreflexive realism, as a means to create correspondence between the mental, formal, and operational labor processes. Nake expresses it as a situation where "signs representing things and signs representing operations merge."[22] In other words, it is a form of interaction that strives toward an ideal mapping of the domains that can transgress the arbitrariness of the sign. Though often building on a deep understanding of the users' perception of the world as expressed in their language, this design approach also uses aesthetics to support the mapping and generate seamless interaction.[23] Andersen has, for instance, studied how aesthetic properties from movies, such as cuts, scenes, or sequences, can be used to design process control in a user interface—among other things, by creating comprehensible relations between sound signals and events, or creating thematic structures in the interaction.[24]

The dream of seamless interaction, embedded in the user's reality, is historically a driving force of much HCI design. Although both Nake and most interaction designers are well aware that seamlessness is an illusion, it is a logic of transparency (to use Bolter and Grusin's term) that persists in images of "Man–Computer Symbiosis" (as expressed by Joseph Licklider in his visions of computer systems), "synthetic telepathy" (such as Edmond Dewans's experiments with mind control in order to send Morse code), and "ubiquitous" or "calm" computing (where information technology leaves the graphical user interface to become part of a tangible environment, as a silent hum in the background, as articulated by Mark Weiser).[25] Indeed, there are endless high-tech myths that address the disappearance of the interface. In other words, interface design often has an embedded prophesy of pure, unnoisy, and seamless signal processes where the problems of representation are overcome in various ways. This has frequently been what fascinates the user, and is also seen today in, for instance, the marketing of the 3D virtual reality headset Oculus Rift or other similar inventions.

Much digital aesthetic theory (including that of Aarseth), software studies, and interface aesthetics can be seen in opposition to this.[26] Rather than developing techniques to overcome the interface, this book regards aesthetics as a reflection of the interface's juxtaposition of sign and signal, or media and instrument. For instance, in electronic literature the assemblage becomes the center of attention, in the sense that it is a literary form that often experiments with what kinds of actions and mediations there potentially could be. In this way, you find literary expressions that reflect the organization of the text, and textual organizations that become important parts of the expression. One example of this (which will be addressed in the book's final chapter) is the *Poetry Machine*, which investigates how readers become involved in the ergodic cybertext production, and thus gain a new role and responsibility for what Aarseth has labeled "the scripton," meaning the strings of text that appear to the reader, while writers become producers of a "textonic" landscape, meaning the strings of text as they exist in the text.[27] Another example (also used in the second chapter) is *The Project Formerly Known as Kindle Forkbomb* by Ubermorgen (2011–2013); it investigates the networked production of text along with publishing models related to YouTube and Amazon.[28]

Electronic literature's reflection on relations between reading and writing, and textural organization, compares well to Bolter and Grusin's description of how visual culture always has had art forms that reflect the particular distribution of the logics of transparency and hypermediacy in a given time. In other words, only in a limited understanding does aesthetics refer to the style and beauty of interfaces. In a critical perspective, it concerns the ways in which the interface reflects new perspectives as well as new ways of perceiving, organizing, and thinking brought about by media technological changes. Fundamentally, as Nake has explained, all these visual, textual, sensual, organizational, and other strategies depend on a particular relation to labor. An aesthetic critique of the interface, and aesthetic interface theory, could begin by understanding this relation to labor as a reading of the artwork's tendency.

Tendency within the Artwork

The aesthetics of artistic practice considered in this book is neither concerned with the beautiful, trendy, or engaging interactive experience, nor a modernistic media specificity (as in Clement Greenberg's understanding of art's autonomy and division into media specific subgenres like oil on canvas, bronze, etc.). Instead, it references a criticism that appears through the material explorations of technology that are found in the arts practices that deal with interfaces, and how they critically reflect the ways they themselves are challenged by the interface and its operations. Or put inversely, if the interface challenges the existing understanding of art, design, and culture, it is because it includes the material and technology as carriers of meaning in a more general perspective. When the artistic practices induce critical reflections on their way of being art and their material basis, they address even deeper political questions of how the interface challenges conceptions of society, politics, and culture. Subsequently, an interface aesthetics and critique is intimately linked with how the artistic interface administers its own technology, and it strives to illuminate how users, more generally through computers, manage the world and produce the reality they are part of. As such, interface aesthetics expresses what the German Jewish critical theorist Walter Benjamin also refers to as a tendency (or *Tendenz* in German) in his seminal essay "The Author as Producer" from 1934.

Tendency is typically associated with a somewhat-banal expression of the commonplace and contemporary, but a tendency can, according to Benjamin, also have a different and deeper meaning related to a material understanding of technology and media. Already in 1927, he writes in a short text that media revolutions lead to fissures in art where deeper tendencies are exposed, and that the art that investigates this—and in this way, uses its medium and technology critically—is the art that contains a deeper political tendency: "But just as deeper rock strata emerge only where the rock is fissured, the deep formation of 'political tendency' likewise reveals itself only in the fissures of art history (and works of art). The technical revolutions are the fracture points of artistic development; it is there that the different political tendencies may be said to come to the surface."[29] Art is in Benjamin's view a probe to the fracture points created by technical revolutions. This is not because artists have a better knowledge of politics than others, but because artistic production is a material exploration of its own technological means of production, and how these constantly change. Consequently, tendency is related to a dialectic material examination of production and technology through artistic production, rather than an abstract ideology or immediate attitude of the work, as Benjamin detects in both fascist and Marxist thinking (the essay was given as a speech in a specific Communist context: the Paris Institute for the Study of Fascism):

> Instead of asking, "What is the attitude of a work to the relations of production of its time? Does it accept them, is it reactionary? Or does it aim at overthrowing them, is it revolutionary?"—instead of this question, or at any rate before it, I would like to propose another. Rather than asking, "What is the attitude of a work to the relations of production of its time?" I would like to ask,

"What is its position *in* them?" This question directly concerns the function the work has within the literary relations of production of its time. It is concerned, in other words, directly with the literary *technique* of works.[30]

Benjamin's highlight of the author's concern for the technique, as opposed to the attitude of the writing, also makes way for a different understanding of tendency than the one found in contemporary autonomist Marxist thinking (e.g., Franco "Bifo" Berardi and Antonio Negri). As noted by Benjamin Noys (in a critique of Berardi), tendency in both autonomous and classical Marxism (e.g., Georg Lukács) points to the "demise of capitalism under the pressure 'of the potency of productive forces.'" Instead of this "apocalyptic" and "eschatological" understanding of tendency, Noys argues (with Fredric Jameson) for a "more nuanced realism about the contemporary conjuncture" where "the aim of critical intellectuals should be to present or represent the contradiction of the time, even to sharpen them ... by practicing a method of the tendency that more closely aligns base and superstructure in our analyses, that permits a closer grasp of the failures, tensions and contradictions of this order."[31] Although Noys does not mention Benjamin, such "ways of practicing a method of the tendency" seem aligned with his call for a preoccupation with a literary technique as a way of not relating to material and political conditions (good, bad, right, wrong, etc.) but instead positioning oneself dialectically *within* them. Without question, "The Author as Producer" is one of the texts that most clearly presents Marxist visions (not to mention given the context in which it was delivered), and it also aims to unravel the eschatological dogmatism of the Marxist tendency toward an analytic, dialectic material exploration, which is nevertheless still political.

In other words, the essay expresses Benjamin's materialistic and technocritical way of thinking about the relation between art, technology, and politics with a focus on literature, and he shows how literature, through its political and literary tendency, can demonstrate general social and technical relations concerning the conditions of production. Its readers should in this way understand the essay's proposal of seeing the author as producer—meaning both a producer of their own text, the apparatus it is produced and published through, and a producer of writing in general. To Benjamin, the author is more than just a writer: "An author who teaches writers nothing teaches no one," he states in his argument for a technocritical dimension to writing.[32] To be an author means to recognize the opportunity that through their work and experiences with changes in a work's own relations of production (i.e., the literary technique), the author can reflect on how production is related to ideology, politics, morality, societal organization, and so on. Literature—and more generally, art—is therefore not just something elevated and spiritual, above the level of the technological and productive, but a kind of technology in itself, too, which is engaged in and part of society. If the author does not recognize this challenge, they will merely be working for the powers that be. Benjamin also cites Bertolt Brecht's critique of the cultural producers who do not understand the apparatus they themselves are part of. Rather than controlling the apparatus of production, they are naively controlled by it. Consequently, art is not only based on a processing of the individual's own existential experiences but also on a concurrent desire to engage oneself in—and even change—the institutions and apparatus of production: the artist needs to act as an engineer. As Benjamin notes, there is in this a decisive difference between merely supplying a productive apparatus and seeking its transformation: "The proposition that to supply a productive apparatus without—to the utmost extent possible—changing it would still be a highly censurable course, even if the material with which it is supplied seemed to be of a revolutionary nature."[33] In its broadest meaning, this productive apparatus encloses not only the author but the reader, too, and the art itself.

Even though "The Author as Producer" primarily refers to authors, Benjamin was widely interested in the new media of his time, such as film and photography, and how they could relate to political realities and the potential for change. Similarly, this book is attempting to cover these underlying tendencies in the arts of interfaces.[34] Technological revolutions—such as the changes seen during the last five decades with the appearance of computers in the military, PCs, supercomputers, tablets, smartphones, Internet of things, and so on—lead in all certainty to fissures within art, and these fissures reflect broader issues in society. Consider, for instance, how the networked distribution of art, music, and literature has challenged the old culture industry as well as the traditional understandings of property, copyright, ownership, and privacy, and how this has led to a new digital culture industry based on streaming and licenses. These new models not only for production but also morals, laws, politics, and much more fundamentally function in an interplay with media technological developments. Moreover, the culture industry is only the precursor of what is happening to other industries where social media not only prescribes social relations but also becomes a platform for marketing, distribution, and a whole economy of sharing (such as Uber or Airbnb), or where open access to data becomes not only a question of public transparency but the basis for innovation and new business opportunities, too (e.g., in urban development, as will be discussed in chapter 3). In relation to work and social life, people generally are organized (in work as well as leisure) markedly different than during the capitalism of industrialization, and the ways value is produced, consumed, and exchanged are undergoing deep changes described by concepts such as, for instance, semantic or semio-capitalism.[35]

The point here is that aesthetics has a further role than simply providing nice design that involves users or facilitates innovative production. The need to engage critically with the kinds of formalizations of labor that the interface implies is related to broader political issues. The interface conceals labor and production processes, including its users' own production of data. Aesthetics is hence a question of exploring the fissures and deeper tendencies of the production conditions, and how the artistic production with interfaces continuously reflects the inadvertent result of the interface's modes of operation (the combination of signs with signals) along with what lies behind, within, ahead of, and beyond them. Informed by both computer semiotics and interface aesthetics, this book's aim is to present a dialectic materialism that can be used in contemporary studies of software and interfaces, and how they produce realities that are entangled with the everyday; the analysis of software's and interfaces' art forms as well as the kinds of wonderful technological imaginaries they may produce; and the innovative and transformative design of the productive apparatus of the interface in a time of metainterfaces.

A contemporary example of an artistic approach relevant to Benjamin's thinking is "The Critical Engineering Manifesto" of the Critical Engineering Working Group (2011–2015). This manifesto argues for a critical engineer who "looks beyond the 'awe of implementation' to determine methods of influence and their specific effects," and "raises awareness that with each technological advance our techno-political literacy is challenged."[36] Echoing Benjamin and the materialist approach present in this book, engineering is considered a form of writing or language that shapes its users, and should be studied and exploited in ways that expose its influence. The people behind the Critical Engineering Working Group are also part of the Weise7 studio in Berlin, and their works are good illustrations of critical engineering. In 2012, they created the *Weise7 book*, which unlike other books, is a wireless independent server that runs from a computer inside the book. By opening the book, "readers" can join the book's wireless network and browse its contents (which includes the images and other material from Weise7's archive).[37] In 2014, they further developed the concept and created a black book of cloth, *The Little Black Book of Wireless*, which includes "a complete history of 802.11 and GSM wireless communications, exploits and vulnerabilities," and simultaneously plays with the reader's own vulnerability by harvesting hostnames, usernames, unencrypted passwords, HTML, and much more. These fragments are published on the book's website, but deleted when closing the book.[38] Through the way they work with technology, Weise7 members explore the tendency of interface production, such as the construction and conflicts of the inscription of reading and writing (this will be addressed in the following chapter).

As Weise7 demonstrates, the interface is undergoing severe changes. The excessive use of "cookies" on the World Wide Web, apps such as Uber, and many more all depend not only on interactions in a self-contained system but on a real-time flow of data between devices and services, too. The signals that users emit along with the capturing of user behavior and consumption patterns become valuable in the sense that they both serve as the impetus for smart metainterface services, and a business model for offering such services for free. In general, streaming services for cultural consumption, social services, and much more function as smart networked services that financially as well as functionally build on the capturing and processing of data—rather than the interface itself as a consumer product that can be bought or sold in a traditional sense.

In other words, the labor and modes of production that the author/producer is engaging with, and the tendency reflected within this production, are changing. The interface, understood as a new form of textuality with other grammars and modes of inscription than traditional text, is entering new contexts that change its material, conceptual, and societal role (e.g., in relation to cultural production, urban settings, and perceptions of climate change or the war on terror, as will be examined in the following chapters). At the same time as the interface is the catalyst for HCI and textual inscriptions of sign and signal processes (i.e., self-contained systems), it is increasingly building on and feeding into a network of interfaces. In this network, the textual inscription does not merely include the formal representation of labor in an algorithmic form (as Nake described) but also an incomprehensible and invisible system that captures and quantifies signals emitted from the use of interfaces, location devices, sensors, and much more. The constant and dissolute exchange and calculation of these data make the signs of the interface appear as a new and smart mode of production. How is one to conceptually understand this change of the interface? And what could the beginning of an aesthetic critique of an interface that conceals itself in significant new ways be?

The Metainterface's Inscriptions of Behaviors

With the metainterface, the computer interface today is seemingly omnipresent. Wherever, there is usually some kind of computer interface available, be it the touch interface of a smartphone or a sensor embedded in the environment. Human–computer interfaces pervade all aspects of everyday life. From being restrained to desktops, they are now everywhere: in pockets and entertainment devices, on walls and buildings, and hidden in infrastructure and environmental sensors. They become smaller and bigger, more mobile and less stationary, than the interfaces of PCs or other workstations.

In this way, the interface has not only become adapted to us by way of technological development but has also internalized a changed perspective on the world, where the online versus off-line, or virtual versus real-life dichotomies, are undermined and hardly make sense anymore in developed societies. The "interface culture," described by Steven Johnson at the end of the millennium, is almost trivial today.[39] This is, for instance, also demonstrated by the US artist Benjamin Grosser's *Touching Software* short film, in which he compiled the numerous times the famous political Washington, DC, drama series *House of Cards* displays interactions with interfaces: characters touching screens, texting each other, reading e-mails, taking photos with their phones, and so on. Touch interfaces are ubiquitous in the series, and constantly intertwined in ordinary activities; in fact, Grosser's edit also documents that interfaces of all kinds become the main plotting device in the many intrigues and conspiracies, and how that fuels our attention as spectators.[40] Yet with the prevalent use of the interface, it becomes masked in new ways that reconfigure interface culture as well.

Notably, the human–computer interface does not merely involve individual clicks on a laptop or desktop screen that address a script in a piece of relatively self-contained software. The ways users experience and interact with computing, and the ways users become inscribed in interfaces by means of reading, writing, and executing the software, are developing beyond the well-known desktop-based graphical user interface. As the interfaces increasingly become mobile, embedded, ubiquitous, and pervasive, they also become part of a network of interfaces whose inner functions are less transparent than the everyday interfaces of cell phones, entertainment devices, and so forth. The interesting aspect is how the everyday human–computer interface is intrinsically linked to a constant flow of signals emitted from users or the environment, and conversely, that it itself emits signals that can be captured and processed. This means that users rarely interact with isolated computers, but instead with networks of data floating in "intelligent" software systems.

On this coded and networked level of interfaces, the everyday interfaces become smart interfaces that make the HCI apt and swift: it becomes a metainterface. This metainterface is based on a constant flow of signals. Examples of these signals range from electromagnetic waves such as a Bluetooth device to other kinds of motions like the transfer of real-time data on the Internet. Even nondigital phenomena such as the weather or flow of traffic in a city emit signals, but in either case signals are always considered to carry value: in order for them to be computed, they must be quantified. In this way, the smartness of the human–computer interface also points to how surroundings are increasingly turned into data that can be used to determine and anticipate events, and how areas that have not been previously thought of as quantifiable are now regularly quantified and computed. Data are captured and stored in so diverse areas as cultural consumption (what people look at or listen to on their devices), cities (from CCTV or the tracking of devices), or weather (rainfall, temperatures, and so forth).

In other words, at the same time as the interface seems to transcend perception, it is omnipresent: it is increasingly difficult to avoid interaction with the networked computers. With this omnipresence, the users' behaviors constantly feed into the smart system of the interface as a cybernetic feedback loop. Moving through traffic with a location service, paying with a credit card, browsing the Internet, communicating via e-mail and social media, reading on an e-reader, watching movies, listening to music via streaming services, playing games online, navigating through web pages with cookies, and so on, are all examples of activities with networked interfaces that themselves emit signals that can be quantified and fed into the system. All the user sees is the metainterface to these displaced processes. This data production is deeply dependent on an increased use of interfaces, such as social media, smartphone apps, search engines, streaming media services, GPS navigators, and the so-called Internet of things, to name only a few instances. It is hard to imagine any "activity-system" that is not thoroughly integrated with distributed computational processes in "smart" and "predictive" ways.

Illustrations of how the everyday and HCI relate to a networked level of *signal-computer-interaction* are numerous. Most people have probably noticed how interfaces become customized in smart ways, and how displays and functions—ranging from search results to route planning—are variables of signals emitted by the user and other users. Broadly speaking, but also fundamentally, the metainterface inscriptions of behavior build on two aspects: data capture and smartness (and techniques of machine learning).

Datafication, or inscription of the world into computational systems, which is implied in the signal–computer interface, has been discussed in various ways previously (though not always in relation to an art context). Data analytics, for instance, is far from a new thing in computing, and many corporations and public institutions have gathered and analyzed data for many years, too. Already in 1994, the information theorist Philip E. Agre discussed how the capture of data is part of "a tradition of applied representational work that has informed organizational practice the world over."[41] Agre argues that "as human activities become intertwined with the mechanisms of computerized tracking, the notion of human interactions with a 'computer'—understood as a discrete, physically localized entity—begins to lose its force; in its place we encounter activity-systems that are thoroughly integrated with distributed computational processes."[42]

To Agre, the use of data capture is not only an extension of human activity through technology but also a structural condition for the use of computers that reflects values and worldviews. In other words, systems of data capture (despite being concealed) change the ways of interacting with interfaces and condition perceptions of the world. This condition is not an indisputable imperative of computational management, but as it is embedded within an everyday use, it is hard to imagine alternatives. Agre, for example, points to the way activities are reorganized in order to become traceable as perhaps the most important effect of datafication. This reorganization involves "representation schemes" that use linguistic metaphors and formal "languages" for representing human activities. "Human activity is thus effectively treated as a kind of language itself."[43] The notion of a grammar of language is traditionally restricted to speech or writing as a system of inscription into alphabetic and phonetic systems that can externalize and communicate thoughts. The externalization of thoughts into systems and procedures, however—or as Nake has explained, a formalization of labor and its formal instrumentalization in the user interface—extends the notion of grammar into human activity. Echoing not Nake but instead the well-known HCI theorists Terry Winnograd and Fernando Flores, Agre describes how this "grammar of action" has a five-stage cycle consisting of analysis, articulation, imposition, social and technical instrumentation, and elaboration.

In other words, already in 1994, Agre sees "the tendency toward ever 'deeper' articulation and capture of activities" as the most significant technical trend, but also insists that it is impossible to "remove the elements of interpretation, strategy, and institutional dynamics."[44] Data might seem raw and immanent to the world and people's activities, but this only happens through initial grammatizing and reorganizing, and such systems are not totalitarian mechanisms but rather leave language for interpretation.

Agre's arguments for a grammar of action is not only restricted to the organization of labor at the workplace. With its widespread use, it is part of the everyday inscription of behaviors that happens with the proliferation of interfaces and grand-scale inclusion of signal–computer interfaces into the human–computer interface. Furthermore, the grammar not only concerns actions that are performed to become traceable (or subject to data mining) but also a grammar built on the imperative of smartness.

The smartness of the networked level of interfaces, generally speaking, relates to processes of machine learning that are concerned with predicting what people want (rather than just knowing what they want). To give an example, services such as Facebook or Amazon offer interfaces that appear as customized to the individual user's needs and preferences. Yet the selection of items in the interface, and statistical models that decide what the user is likely to like, are not just the results of tracking the user's own preferences and interactions (friends, likes, purchases, browsing, scrolling behavior, etc.) but also the massive data capture of many users' behaviors along with complex algorithms that compare them. As pointed out by Chun, this movement from analysis to prediction builds (among other things) on collaborative filtering algorithms that create "neighborhoods" based on similarities and differences between user responses to recommendations.[45] This means that user data are not just kept as a record of an individual customer that the system may respond intelligently to but instead gain value by their contribution to a large statistical body that the user responds to.[46] Hence, in these smart systems, the human is not the only standard for intelligence (as often imagined in theories of HCI). The kinds of intelligence that machine learning processes imply are different from the artificial intelligence of expert systems, as they were envisioned in the 1980s, and the Turing test (testing the ability to distinguish between the actions of humans and computers) does not seem to comply any longer, or rather, the Turing test only complies insofar as the human is acting more like the machine.[47]

In addition to being smart, predictor systems are also all encompassing, and include all aspects of the world. As the science and technology sociologist Adrian Mackenzie points out in his elaboration of machine learning processes, machine learning, like data capture, has a long history within computer science dating back to the 1960s, mostly in specific settings such as credit risk assessment. Today's techniques include decision trees, neural networks, association rules, k-nearest neighbors, expectation maximization, latent semantic analysis, and much more.[48] The general idea is known from statistics' linear modeling of the relationship between dependent variables (such as the price one is willing to pay for an item) and independent explanatory variables (such as income, gender, number of children, etc.). As Mackenzie explains, classical statistics is limited in its predictions because of the difficulty in calculating large matrices of numbers, and it typically depends on a few variables such as gender, age, income, or occupation. By contrast, today almost anything can be considered a variable, and almost anything is juxtaposed and associated in the statistical regression models that form the consumers' or users' neighborhoods. The predictive models of machine learning explore high-dimensional patterns rather than just associations between limited variables.[49]

Not unlike Agre's hypothesis that data capture is a language with a grammar, Mackenzie notices that predictive modeling gives rise to new forms of agency, and "progressively interpolates and interpellates subjects."[50] Similarly, Chun has argued that neighborhood predictors dissolve a postmodern disorientation and subject incapable of apprehending the multiple relations that construct their identity. Instead, an actor replaces the subject, and the subject's narrative is replaced by the capturing of actions.[51] This change in formality is also visible to the user, and seen, say, in the ways maps and dashboards (graphs, analytics, etc.) have replaced the subject and three-dimensional imagery as an origin of perspective in interface design. In other words, interface design signifies the ways individuals, objects, and their relations are processed by neighborhoods and other predictors, and inscribe the user in a grammatical system. Subject formation is an outcome of a complex process that inscribes "actors" and "actions" in the interface, and where the smart processing of actors and actions is as important as their traceability.

Like data capture, the inscriptions of behaviors that take place in machine learning processes participate in world construction as well as, for instance, the constitution of the subject's identity. Yet the relation between processes of individuation and inscription is naturally also a complex philosophical issue that has been brought up by, for one, Jacques Derrida, who describes it as an exteriorization of consciousness and memory.[52] Both Chun and Mackenzie seem to suggest that this externalization, which lies in the inscription's transformation of thought and memory into discrete units, depends on technical systems that are undergoing severe changes. The constructions of writing (called "grammatology" in Derrida's terms) is not just dependent on a formal alphabetic system with a grammar but also on material constructions, or textual machines, that run on alphabetic code (as formerly explained), data capture, and smart algorithms. Subsequently, interfaces, and their encouragement to compulsively "like," "retweet," or in other ways inscribe their actors and actions in a smart system, are not innocent activities, nor are they extensions of human activities. Instead, they inscribe the human in a system with a grammar where they are not always in control or conscious of writing but nevertheless can leave signs for human interpretation and critique.

The grammars of these inscriptions that take place through the combination of HCI with signal–computer interaction (i.e., data capture and predictive smart modeling) have a certain persistence. Antoinette Rouvroy, a doctor of law at Namur University, has argued that "data-behaviorism" (as she labels the grammar) and its "statistical inference operated on the basis of correlations" is validated "through a kind of 'backward performativity': anything that would happen and be recorded ... will contribute to the refinement and improvement of the 'statistical body.'"[53] When data regarding consumption, transactions, social relations, and so forth, becomes indicative of the reality to come (which cannot yet be felt), it has an impact on how technical infrastructures are constructed. The wide proliferation of interfaces encourages the users to act. This may happen consciously as when they willingly express their preferences on

Table 1.1
Elements of the human-computer interface and the signal-computer interface

Human–Computer Interface	Signal–Computer Interface
Browsing, scrolling behavior, likes, reviews, posts, and other user responses to the metainterface Sensors, cookies, Bluetooth, etc.	Techniques for tracking behavior that contributes to the statistical body (data capture)
Recommendations, requests, posts, routes, and other experiences of personalization and customization	Machine learning and collaborative filtering of data (smartness)

social media, but also unconsciously when they are tracked by cookies or positioning systems in their devices. These actions and behaviors in the metainterfaces (platforms, urban interfaces, and more) are thus part of a signal–computer interface and contribute to the statistical body that forms the basis for the users' interfaces in a feedback loop (table 1.1). Users, however, also dislike, avoid suggestions, take different routes, or in other ways misbehave. Such disruptions of the loop do not dismiss the modeling and its grammar; they merely cause an adjustment. They may contribute to the formation of new neighborhoods and are even actively encouraged by the systems.[54] The system, its processes of interpellation, and so forth, are still in place, and continually becoming ever more differentiated. This strange temporal feedback loop is what creates the product of anticipation in the metainterface.

Interface Criticism

Clearly, many things are at stake in the metainterface when the human–computer interface gets coupled with a signal–computer interface. Hence, the interface's inscriptions of behavior through data capture and machine learning can be critiqued in many different ways. Generally speaking, they can be critiqued at the levels of ethics, logics, and aesthetics.

Evidently, it is not difficult to question their validity. To many, the excessive data capture should not be officially and legally acceptable, and it is also common knowledge that it takes place on grounds that are opaque and not always entirely honest. This goes for the wide use of cookies on the World Wide Web to the tracking of locations and user data by apps. As smart as interfaces may seem, they obviously also have their logical limit: they can be difficult to use, or produce outcomes that are useless and simply not rational. It is hard to make up for human experience and easy to put users in the wrong "neighborhood," and the smartness of Google, Netflix, Amazon, and others may turn out to be a well-programmed presentation of things users have a hard time grasping or do not really want. Critique may also be of a different kind, however. Rather than the validity and logic of data capture and predictive modeling, one may address their operationality as such, and how the textual machinery not only is a functional entity but also a kind of linguistic formalization of labor with a particular grammar that is veiled, though nevertheless possible to address. This knowledge is the aim of aesthetics. But where does an aesthetic critique begin?

As previously discussed, artistic production with interfaces has traditionally been concerned with an interface that disappears behind a veil of immersive and transparent actions that generally speaking, not only convey meaning but meaning and action simultaneously, too. This juxtaposition of processes of signification with signal processes is often unveiled as a language with specific properties that refer to a system of production. By reflecting its own textual production of inscription, as an encoded textual machinery, the artwork (and "author as a producer" of interfaces) potentially reflects the larger conditions of production along with their influence on cultural values such as reading, writing, sharing, producing, and so on.

The interface has seemingly functioned as an intangible window into an alternative, virtual reality that could be operated with buttons, menus, pointers, and so forth, and this operational realism and reality effect of the interface has been countered by interface arts that in different ways have reflected its operationality as a particular mode of production. Put differently, the artistic productions have functioned as realism in its literary sense, reflecting the visible/invisible procedures of the interface as a tendency rather than producing reality effects.[55]

When it comes to the metainterface, the coupling of human–computer interfaces with signal–computer interfaces, and the textual production and processes of inscription involved in data capture and the smartness of machine learning, this realism and tendency in the artwork is of a different kind. As Rouvroy has argued, "Raw data function as de-territorialized signals, inducing reflex responses in computer systems, rather than as signs carrying meanings and requiring interpretation."[56] In other words, in the coupling of human–computer interfaces with signal–computer interfaces, knowledge is seemingly not produced about the world anymore but instead from the world. The epistemic knowledge produced in the metainterface (and that is intrinsic to the intelligence of contemporary interface culture) has lost its contact with the world it is aimed at representing, yet it simultaneously presents itself as an ontology. This means that the substitution of interpretation with computed and automated reflex responses is not an overcoming of a problem of representation. Whereas the design of human–computer interface traditionally attempts to overcome the interface by displacing its written and constructed character, the coupling to the signal–computer interface in the metainterface transforms the entire world into (the production of) writing or data, but displaces its own grammar of selection, processing, and inscription: on top of the illusionism and transparency of traditional interface realism, a data realism is added. In their realism and way of relating to the real, the metainterfaces of search engines, weather apps, location services, and so on, are not just seamless representations of *virtual* realities but rather *augmented* ones. They are no longer just windows to other worlds that can be manipulated; instead, they appear to be enhancements of reality. These enhancements are the result of intelligent processes that capture and process data, and are presented to the user as overviews that filter, quantify, update, and prompt news, traffic, messages, advertisements, playlists, social relations, and the like.

Following this, the new challenge for an aesthetics of the metainterface is not only to discuss the fusion of physical and virtual reality as an aesthetics inscribed in the human–computer-interface (control and media, opaqueness and transparency, etc.) but instead to address this concealed production of reality in the metainterface: how it reappears and leaves signs of its own process of inscription. Interface aesthetics implies an ability to question how signals in the form of captured data (supposedly immanent in the world) and their intelligent processing are inscribed into the world as language with a grammar that affects perceptions of reading and writing, spatial properties, and even sense perception (as the following chapters will address). The metainterface is a new production mode that increasingly transgresses the relatively closed system of self-contained software and HCI. It incorporates a signal–computer interface that quantifies and datafies, and ultimately turns the whole world into an interface: a large statistical body whose reality deeply depends on the processing and visualizations of data.

One example of a practice that addresses the concealed production of reality in the metainterface is found in the works of Ben Grosser, and also reflected in his own writings. Through the removal of all metrics on Facebook's interface (numbers of likes, friends, shares, comments, requests, etc.), the browser extension *Facebook Demetricator* (figure 1.1) is an exploration of how the seemingly innocent and smooth social interface leaves signs of its own process of inscription as well as contains particular grammars of action.[57] The extension lets the users experience the effects on their own social individuation processes on Facebook. In Grosser's own words, such processes depend on what he labels a "graphopticon": a "form of self-induced audit within social networks like Facebook where the many watch the metrics of the many." As such, graphopticons—the overwhelming elements of metrification on the Facebook interface—is a parallel to Agre's grammars of actions, yet Grosser's work demonstrates how the grammar is produced through concrete interface elements that relate to the quantification of individual and social behavior.[58] In fact, it points out exactly why quantification elements

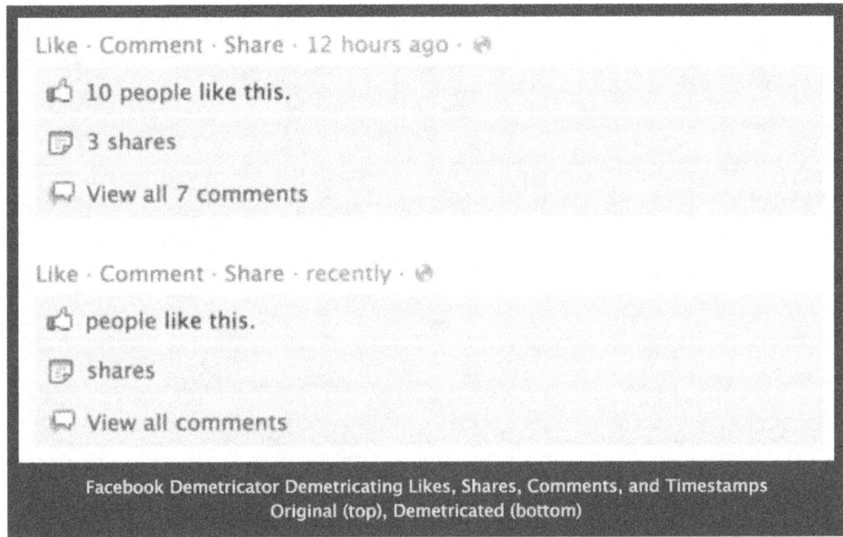

Figure 1.1
Facebook Demetricator (2012–) by Ben Grosser. Browser extension that allows the user to experience Facebook without quantifications such as time stamps, number of likes, requests, etc. Courtesy of Ben Grosser. Screen capture.

become a visible interface or graphopticon, and why metrics is not always hidden as an invisible part of data capture and smart algorithms. Given this, *Facebook Demetricator* demonstrates a tendency: "Through its metrics, Facebook imposes patterns of interaction on us, changing what we say to each other and guiding how we think about each other. Demetricator, through its removal of the metrics, both reveals and eases these prescribed patterns of sociality. It shows us what the metrics want. The metrics want *more*."[59] It is the logic of quantification, the wish for "more" that makes it necessarily visible as an interface element users see and react to—a grammar of action that entices users to participate in the quantification of the social and self.[60]

The following three chapters track the development of the metainterface through artistic productions and practices that deal with, respectively, the platforms that convey interfaces for cultural consumption (e-readers, smartphones, and tablets), spatial inscriptions of urban interfaces, and clouds of today's cultural interfaces. Through analysis of specific artworks and their tendency, they discuss how these new inscriptions and grammars of the metainterface appear at the linguistic formal, spatial, and material levels, and eventually how to design with this critical awareness, as addressed in the final chapter.

2 The Metainterface Industry: New Platforms for Culture

RealBeat

RealBeat is an iOS app by the Austrian media artist and poet Jörg Piringer (figure 2.1). It is a small sampler and sequencer made for the iPhone and iPad that lets the user record and transform everyday sounds into musical compositions. *RealBeat* transforms the iPhone into a mobile recording studio, and demonstrates the potential of the smartphone as at once a recorder, studio with effects, live processor, and playback device. By allowing users to easily record, transform, and compose with the sounds of their everyday environment, it points to how the metainterface of the smartphone potentially changes our way of relating to and being in our (sonic) environment, and makes it relatively easy to compose beat-driven electronic music that is otherwise quite technical.

In several ways, *RealBeat* utilizes the device to fulfill many of the twentieth-century aesthetic sonic visions. Already in 1913, the futurist Luigi Russolo argued that we should use the noise of the city as an instrument in music, the 1940s' and 1950s' musique concrète developed the recording and composition of everyday sound, and in the 1960s and 1970s, R. Murray Schafer and other composers maintained that we need to care about and listen to the soundscape and acoustic ecology of the environment.[1] Also in popular music, recorded sound, sampling, sequencing, and noise have become part of a musical repertoire. Clearly, *RealBeat* is an example of how musical aesthetics, once reserved for the avant-garde and owners of expensive studios and equipment, has become a cheap, commonplace accessory, readily available in our pockets. It demonstrates how smartphones potentially transform experience into an active recording, sampling, and manipulating activity while waiting for the bus or cycling along in the city. It allows for a new kind of active listening as a way of fulfilling Russolo's prediction that in the future, we would be able to combine "thirty thousand different noises ... according to our artistic fantasy."[2] As such, *RealBeat* follows a trajectory seen within visual culture, too, where mobile cameras and social media have allowed their users to

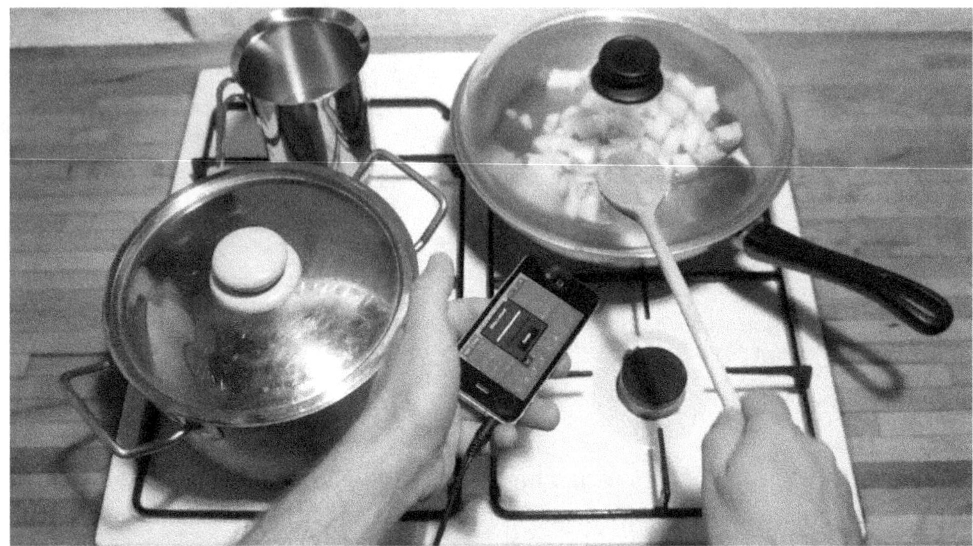

Figure 2.1
RealBeat (2011) by Jörg Piringer. A sampler and sequencer made for the iPhone and iPad that lets the user record and transform everyday sounds into musical compositions. Courtesy of Jörg Piringer. Screen shot from instruction video.

photograph, filter, manipulate, or in other ways interact with the immediate environment, and distribute their daily experiences.

RealBeat is an example of how a software-based art practice, with clear references to an art historical context, has eventually found a stable platform for production, distribution, and consumption. One may even argue that the sonic practice presented in *RealBeat* has become mainstream. Given that the app-based platforms also feature streaming services for music, books, and movies as well as numerous social media apps along with thousands of games, it is apparent that these platforms—which include smartphones, tablets, watches, e-readers, TV and music streamers from companies such as Apple, Google, and Amazon, and so on—represent a cultural turn in computing. From the avant-garde to the mainstream, software-based aesthetic production and distribution seems to drive the market for the metainterface and its platform computing. Or put differently, the platforms have become a new conveyor belt for the networked computer as a cultural industry and the functional basis of the metainterface's production system.

Besides a steady flow of games, most of the content for cultural devices are digital versions of more or less traditional cultural content where the digitization seemingly does not influence the aesthetic form, medium, or content. Typical illustrations of this are streaming music services, streaming film and TV services, and e-books. Yet the metainterface industry is also an opportunity for more innovative aesthetics. Within electronic literature, a number of projects have related to both smartphones and tablets. Interesting example are Jason Edward Lewis's and Piringer's poetry apps that use the touch screen interface to allow readers to "touch" letters and words in concrete poetry universes, or narrative apps like Danny Cannizzaro and Samantha Gorman's *Pry* from 2014. Also within art and music, a rich variety of apps critically explore all aspects of the new platforms for interface culture along with their discreet networks, displays, location systems, cameras, accelerometers, microphone, and much more. The metainterface industry is obviously an opportunity for software artists—with a stable distribution system, and even a business model that allows content creators to create a small income (after giving Apple or Google their share), but it also presents challenges and problems for both art and computing.

This chapter focuses not only on how computing turns cultural with the metainterface but also how this implicitly involves a computation of culture with automated and software-driven means that, for instance, automatize the process of curation. Seemingly the new platforms empower both user and producer. As it will be argued, however, the new model for cultural production, distribution, and consumption that comes with the metainterface embeds technocratic protocols that disguise particular perspectives on culture. More specifically, the chapter looks at how consumption becomes heavily monitored and datafied, and hence also a new mode of production in a metainterface industry. The aim of the chapter, in other words, is to provide a characterization of this new industry, reflect on how it relates to former culture industries, and envisage how cracks in the system allow for new imaginations as a tendency within the cultural expression. For instance, the chapter will explore how particular apps and projects innovatively expose, utilize, and subvert the metainterface industry's inherent procedures and restrictions to exemplify its ambiguities, and offer imaginary and concrete alternatives in a legal gray zone, too.

In the Office

The present cultural turn in computing rests on an awareness of how the often-bureaucratic and office-related procedures of computing constitute a new language, and thereby also lead to new cultural expressions. In computing, this aspect has historically received little attention, but is nevertheless essential in an understanding of the present cultural turn.

Throughout the history of human–computer interaction (HCI), the preference in interface design has always been user-friendly office work. In the 1980s and 1990s, computing was seen as a revolution in office work. The operating systems of first Apple (Mac OS, 1984) and later Microsoft (Windows, 1985) was structured around the desktop metaphor, with folders, documents, and so forth, and PCs were widely marketed through software packages for office work, such as Microsoft's Office package (1988) with word processing, spread sheets, and slide shows. As Warren Sack states, "[That] these operations correlate almost exactly with what the bureaucrat does with his file cabinets, desk, and trash can is no coincidence." It is therefore "clear that shuffling through, stacking, listing, and filing were the ideals of 'memory' and 'thought' admired and implemented by the founders of computer science and interface design."[3] In a larger genealogy, these ideals also originate from the military system that developed modern computing and programming. This has been elaborated by Wendy Chun, who also argues that the very programmability of the computer depends on an ability to store programs and information in the computer's memory, and that its genealogy exposes an idiosyncratic image of memory as storable.[4]

In HCI, there has historically been only a limited interest in how computing has developed as a cultural phenomenon. If at all, the interest has been in how the field could learn from, for instance, computer games. As Ben Shneiderman, a key figure in HCI, wrote in the early 1980s, where video games had experienced an immense popularity, games are "perhaps the most exciting, well-engineered—certainly, the most successful—application of direct manipulation." Joysticks, buttons, and other physical actions contain principles that "can be applied to office automation, personal computing, and other interactive environments."[5] The present cultural turn in computing, however, differs from Shneiderman's pragmatist interest in user-friendliness that wants to do away with the syntax errors of the command line interface and the like so as to create a more human-centered interface design. Rather than seeing the interface as something disappearing in the use situation, the turn is founded on the language of the interface—in its grammars of action, and aesthetic and cultural implications. This language is, for instance, brought forward by Matthew Fuller in his seminal article "It Looks Like You're Writing a Letter: Microsoft Word"; Microsoft Office's design is not only a useful design for office work but also a design of a user, and particular bureaucratic and businesslike ways of being active and laborious.[6] In other words, the software interface contains a range of predefined functions whose purpose is not simply to empower the user (as imagined by Shneiderman, and others) but also to regulate the user's writing and render them "a subject under the word processer" who is not just writing texts but *Word documents*, too, and thereby creating a "de facto standard for word processing much to the handicap of other word processors."[7] Or as Geoff Cox states, "The writer becomes part of the machine, thoroughly embedded in the choice of computer and software program."[8] To understand how software and interface become cultural, one needs to understand this compliance between the author and the bureaucratic machinery of the computer, including the ambiguities of this relation. The cultural turn is then not just a history of pragmatist assimilation but also the creative development of uses of the software and cultures of writing, as demonstrated by Matthew Kirschenbaum in his *Track Changes: A Literary History of Word Processing*."[9]

Both Fuller and Piringer have been active in producing cultural software that explores the operationality of the computer, or software as an aesthetic phenomenon. In relation to Microsoft Word, Fuller suggests that users, for example, "cut the word up, open, and into process."[10] In his own installation *A Song for Occupations*, for which his text on Microsoft Word was originally conceived, he pedantically maps the software "to take apart a component of this body, that part that 'produces' writing, and to find out how it is composed as a sensorium, how it machines language," as he explains.[11] Piringer's earlier work, the visual poetry generator *nam shub*, can be seen as taking up this challenge.[12] If *A Song for Occupations* and *nam shub* explore the sensorium of machinic language, *RealBeat* is a sonic practice that follows the same avant-garde tradition of cutting up, opening up, and putting into process.

In the present cultural turn in computing, though, one may argue that such aesthetic practices not only have become mainstream but also subjugated. Apps for sampling, editing, and displaying recordings of everyday experiences are among the most popular in Apple's App Store, though with a preference for visual rather than sonic recordings (popular apps such as Afterlight or Facetune are good examples of this).[13] In the inclusion of cultural computing, the App Store and other similar phenomena also significantly change and bureaucratize the possible compliance between the author and the machinery of the computer.

In the Office of Culture

If user-friendliness in office-based interface design brought the computer to the masses in the 1990s, the current cultural turn in computing presents software interfaces that are cultural by default. Like sound has music, and writing has literature, software has its own cultural expressions. Whereas books, music, movies, and so on, have had solid platforms for production, distribution, and consumption (publishers, stores, libraries, etc.), though, cultural software has not had its platforms until now. With the metainterface (i.e., Apple's App Store, Google Play, Amazon, and other services), producers of cultural software now have an interface to the new consumers of computer culture—with a well-defined business model and on hardware platforms readily available to billions of users. If the 1980s and 1990s were dominated by the PC interface, the 2000s have become dominated by cultural computing: devices, software, and interfaces related to cultural content and social media are constantly developed, updated, and integrated into everyday spaces, interests, and tasks.

Mainstream interface design's interest in computing and software as a cultural phenomenon is obviously not restricted to experimental software on the iPhone. With streaming services for music, movies, audiobooks and e-books, games, social media, and many other apps, it is evident that the metainterface is a new culture industry. Yet as will be discussed in more depth later in this chapter, this new culture industry differs in several ways from the industry of mass media. The easy access to apps and networks provided by the new computer platforms of the metainterface increases the consumption of culture (music, video, games, text, etc.), but implies that independent and niche productions (such as Piringer's *RealBeat*) can reach a dispersed audience (also framed by Chris Anderson as the theory of the "Long Tail").[14] In other words, the market for culture is changing dramatically, and in ways that benefit both mainstream and niche. Though there are of course both winners and losers (including content producers that get a tiny revenue), it still seems that "the sky is rising" according to a widely referred US technology report: people are generally increasing their spending on culture.[15]

Any kind of cultural critique of metainterfaces and their industry must include not only a general critique of the interface but also a critique of the general user-friendliness of the interface—meaning a critique of the bureaucratization of cultural software that takes place in the metainterface. In contemporary interface culture, cultural production is tailored to the distribution platforms, and to a large extend becomes "shrink-wrapped" and fitted—like office work—into particular formats and predefined platforms. In this way, cultural production and consumption become connected as input and output in the same metainterface, often blurring the differences. Reading books, listening to music, or watching movies have hitherto been considered a passive engagement, but is now an invaluable part of a chain of production: the successful prediction of what people will consume deeply depends on processes of monitoring, quantifying, and calculating consumption in controlled environments that can predict general behaviors. As behavioral monitoring, profiling, and production of predictions work behind the simulated pages of e-books or streaming of music, the human–computer interfaces merge with signal–computer interfaces in the metainterface, and the functional basis of this is the creation of a new cultural metainterface industry.

Hence, just as much as computing has turned cultural, culture is increasingly computed and bureaucratized. The new culture industry has, in other words, subsumed the kinds of software that explore the aesthetic and cultural implications of software. Seemingly, the aimless cultural occupation of messing with text, sound, and images has beaten office work. Yet this turn also implies that "office work"—and its bureaucratic procedures of monitoring, storing, reporting, and so on—has now become applied to cultural production and consumption. The scale of this, and how the cultural computer is used to compute and predict cultural, social, and individual behavior, is exorbitant. The implications of monitoring user behavior and other forms of datafication is, on the one hand, a sense of "ease of use" where networks of other computers and users serve cultural products intelligently on a plate, and on the other, a sense of discomfort.

As an echo of Fuller's plea to address the use of office software design critically, the aim must be to cut up and take apart the machine that produces and distributes writing, or any other kind of cultural expression, and explore how the metainterface functions as a sensorium that "machines" language and designs its consumers—or in Walter Benjamin's term, explore its tendency. The challenge that Fuller proposes is not only analytic but also practical. Whereas *A Song for Occupations* and *nam shub* reflect the operations of word processing and a work-based *office industry*, the aim is to present how a practice may relate to a cultural *metainterface industry*, which automatizes, bureaucratizes, and instrumentalizes cultural production, distribution, and consumption. The following parts of this chapter will explore the new cultural industry of metainterfaces, first by reflecting on platforms of the metainterface and their control mechanisms, then on how the platforms reconfigure everyday life (e.g., produces new users along with new forms of writing and reading), and finally on how this industry both resembles and distinguishes itself from a former culture industry.

Software Art Platforms—from Curating to Autocurating

To begin an exploration of how platform control is exercised in the metainterface, it is fruitful to return to a situation before the platforms settled. *RealBeat* is an example of how software art and aesthetic explorations of interfaces no longer are restricted to repositories such as runme.org, where Piringer previously has advertised his visual poetry generator *nam shub*.

Runme.org (figure 2.2) was launched in 2003 and developed with strong ties to the Readme festivals, organized by the net and software artist Alexei Shulgin, Olga Goriunova, and others. The Readme festivals consisted of four consecutive events in Moscow (Readme 1.2, 2002), Helsinki (Readme 2.3, 2003), Aarhus (Readme 3.4, 2004),

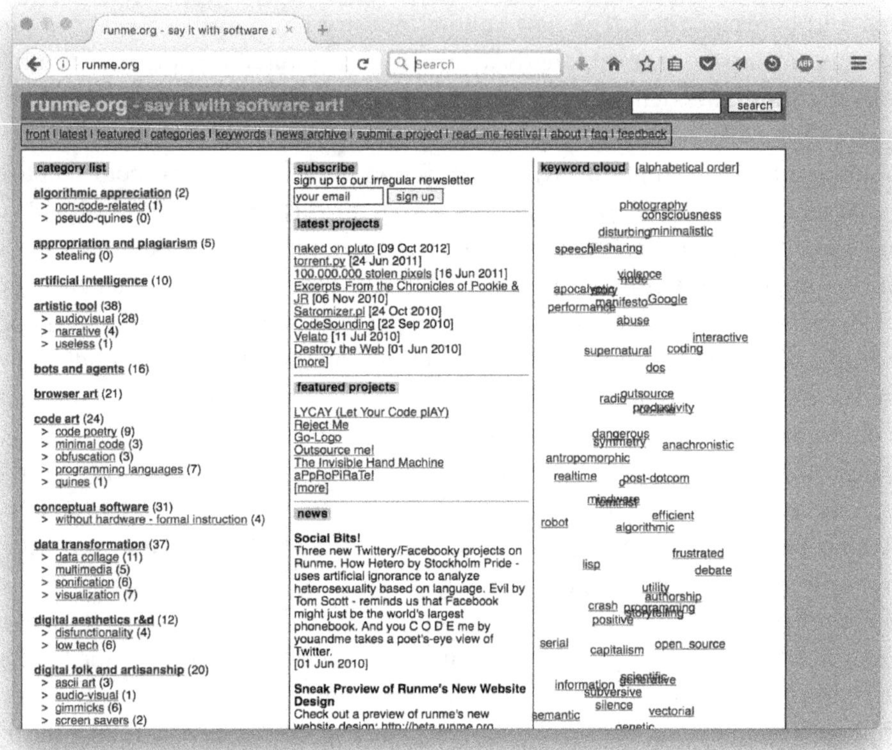

Figure 2.2
Runme.org (2003–) by Amy Alexander, Olga Goriunova, Alex McLean, and Alexei Shulgin. An online and open repository where everyone can submit examples of software art. All entries appear in the database (in lists, keyword clouds, etc.)—some are also featured with expert reviews. Screen shot from website.

and Dortmund (Readme 100, 2005).[16] Already in 1997, Shulgin had launched a "Form Art Competition" for the big Austrian festival for electronic art, Ars Electronica, that hitherto had been focused on art produced with computers and electronics, and less on computers and software as expressions in themselves. The submissions featured drop-down menus, radio buttons, text fields, and window frames as interactive images—turning the otherwise-functional metaphors for interaction on the World Wide Web into conceptual art. The Readme festivals served as platforms for the critical research of software as a cultural phenomenon rather than a mere tool for solving tasks. In other words, their aim was explicitly to frame the field of ongoing artistic activities with software (i.e., software art), and build a theoretical framework around them that included scholars from aesthetics, sociology of technology, literary theory, cultural studies, philosophy, and so on. As such, they appeared as part of a larger trend that included the Berlin-based Transmediale Festival's software art award, art.bit in Tokyo, Liverpool Biennal's Generator, Whitney Museum's code.doc in New York, Electrohype in Malmö, and the 2003 edition of Ars Electronica that featured "CODE" as its main topic. At the time, they represented rare opportunities for scholars to gather, and an audience to access and experience software art.

Whereas other cultural expressions (like literature, music, movies, etc.) had solid platforms for production, distribution, consumption, and critique (e.g., literature has had a print industry with publishers, libraries, stores, and professional critics), it was significant that platforms for cultural software other than games were only known in narrow circles, and were largely collective, self-organizing, and in flux. Yet at the same time, they sought to mime the procedures of an art field, such as festivals and critical peer reviews. As stated on runme.org,

> Software art gets its lifeblood and its techniques from living software culture and represents approaches and strategies similar to those used in the art world. Software culture lives on the Internet and is often presented through special sites called software repositories. Art is traditionally presented in festivals and exhibitions. Software art on the one hand brings software culture into the art field, but on the other hand it extends art beyond institutions.[17]

In the spirit of Internet culture, runme.org is an open database where any user can submit projects they consider interesting examples of software art. This freedom, inherent to a free and open-source software culture more generally, is followed by a particular mode of regulation known from the art world's curatorial practices. Though the invitation to submit is open, and all entries appear in a keyword cloud, they are, for example, also listed in predefined categories, and only some of them will get reviewed as recognition of their quality. These reviewers (Amy Alexander, Florian Cramer, Fuller, Goriunova, Thomax Kaulmann, Alex McLean, Pit Schultz, Shulgin, the Yes Men, Hans Bernhard, and Alessandro Ludovico) were part of the community and functioned—with a term from an arts context—as "curators."

With the metainterface, in the new interface industry, this relation between content production and curation no longer implies a peer relationship, and this considerably changes the nature of the platform. The cultural industries have had a hard time adapting to a living software culture of rips and remixes (which *RealBeat* alludes to as well), but the platforms seem to offer new and viable solutions for cultural production, distribution, and consumption. As much as they allow for a new interface-based cultural industry to flourish, however, they also inflict censorship and control. Although runme.org exercises control as a form of curating, the curating of the platforms of a new metainterface industry differs significantly from this. If one compares App Store's curation of *RealBeat* to the kinds of curating on Piringer's *nam shub*, submitted to runme.org, it becomes evident how cultural critique increasingly becomes administrative. Although it has never been featured and reviewed by peers, *nam shub* is readily identifiable and available as a software art project on runme.org. Contrary to *nam shub*, *RealBeat* has been reviewed and found suitable for the cultural platform. Yet instead of employing cultural experts, Apple has strict protocols for the review process, which allows for automated and large-scale reviews, including not only bureaucrats but also software analysis and user report systems. In this sense, the new cultural platforms implicitly render expert critique abundant, and Apple has neither category nor appreciation for art in its App Store.

Examples of how Apple's assessment exceeds the app's functionality are numerous, and most often follow rules that are not always explicit but instead hidden within the system as well as its software and hardware. For instance, apps need to serve a clear purpose; they need to submit themselves to a strict moral codex that excludes the use of explicit lyrics, visuals, and so on, and they need to follow the codex of copyright.[18] All these protocols by definition rule out most cultural and artistic practices. Artistic practice does not have a clear purpose; it may also involve explicit expressions, and software art often violates copyright. If sampler-based music was (in)famous for the ways it violated copyrights and reused musical tracks, it is therefore remarkable that *RealBeat* does not provide access to the user's music library. There might be artistic reasons for this, as Piringer's artistic production is characterized by an interest in concrete sound and poetry, and he is also part of the Vegetable Orchestra (which performs on self-made instruments made out of vegetables). Yet there is a technical reason for this, too. As Apple states on its development website, "Apps that do not use the MediaPlayer framework to access media in the Music Library will be rejected."[19] Or put differently, "Every App is an Island," and any access to the system directory is off-limits to the developer.[20]

There are numerous stories about how apps have been delayed or rejected by Apple's gatekeeping because they violate the developer guidelines (their functionality is misleading, they display nudity, they have a "wrong" business model. etc.).[21] Apple has, like Facebook and other companies in the metainterface industry, extremely vague guidelines. As it states on its review guidelines site,

We will reject Apps for any content or behavior that we believe is over the line. What line, you ask? Well, as a Supreme Court Justice once said, "I'll know it when I see it." And we think that you will also know it when you cross it. ... We view Apps different than books or songs, which we do not curate. If you want to criticize a religion, write a book. If you want to describe sex, write a book or a song, or create a medical App. ... If you run to the press and trash us, it never helps.[22]

This can only be read as a call for self-censorship, an explicit evasion of the freedom of expression, which even warns against raising a public debate. It reveals that Apple's curatorial control is not restricted to technical issues (bugs, use of file formats, human interface guidelines, etc.) but also includes larger cultural debates, even though it shows a lack of understanding of critical, artistic content. Unlike the major record labels, publishers, and so forth, in the former culture industry, it has monopolistic control over the market in its App Store: if an iOS app is rejected, it is cut off from mainstream distribution to Apple's devices.

Phone Story (figure 2.3), from the independent game producer Paolo Pedercini and his company Molleindustria, perfectly illustrates how apps that critically reflect the cultural and political reality of their own production—that is, their *tendency*—may end up in trouble when faced with Apple's curatorial control. *Phone Story* is a game made for Android and iOS that "attempts to provoke a critical reflection on its own technological platform."[23] It does so by letting the user play small episodes that demonstrate the production and marketing of smartphones, including the exploitative mining of coltan in Congo, suicidal conditions of sweatshop workers in China, manufacturing of a constant desire for new products through obsolescence, and problems of e-waste when the old products are discarded. The game's website elaborates how the overall idea of the game is to collect money for nongovernmental organizations that fight corporate abuse. The game, however, was only allowed in the App Store for a few hours, and less than a thousand users managed to buy *Phone Story* before Apple decided to withdraw it.[24] Though Apple has rejected *Phone Story* for various reasons, including its way of generating money for charitable organizations, it is hard not to see it as one example among many of how Apple does not tolerate criticism of its platform in the App Store. Consequently, it is difficult to expose and critique the cultural and political reality of the metainterface industry from within its own platforms.

In other words, the automated protocols for curating cultural software—and indeed any media distributed on the platforms—reflect a particular control of cultural consumption (the enforcement of copyright, protection of brands and business models, etc.) that is embedded within the technical infrastructures of the prevailing cultural platforms of the metainterface: the smartphones, phablets, tablets, and e-readers along with their controlled environments of "stores" and "apps."[25]

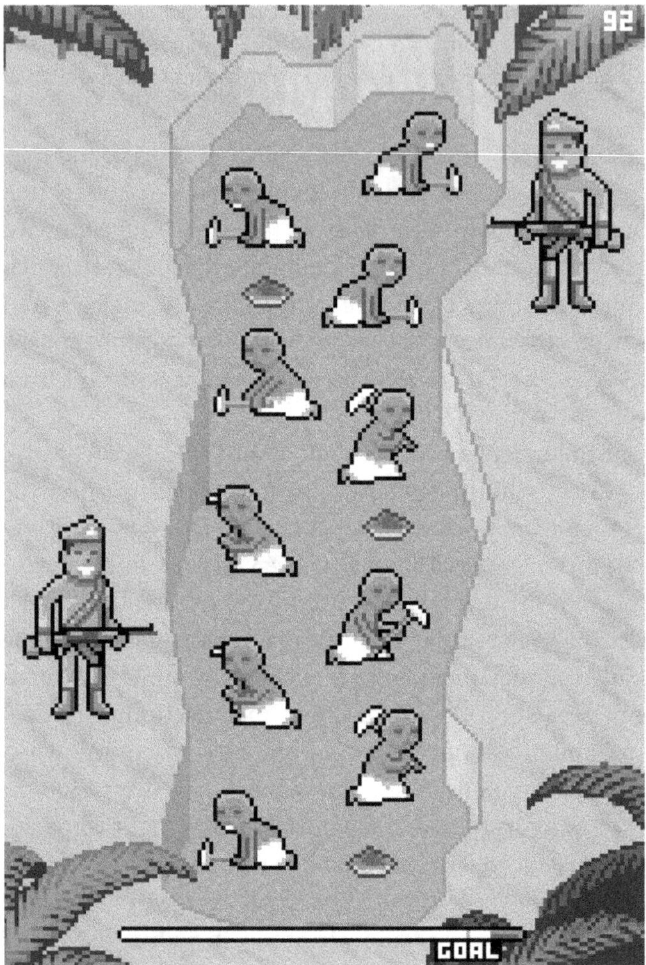

Figure 2.3
Phone Story (2011) by Molleindustria. A game made for Android and iOS that lets the user play small episodes that demonstrate the production of smart phones, including the exploitative mining of coltan in Congo (illustration). The game was banned from Apple's App Store. Courtesy of Molleindustria. Screen shot from app.

The Control of Production, Distribution, and Consumption

The kinds of bureaucratic and automated control mechanisms that are involved in the large-scale curating of cultural content in, for instance, the App Store reflects a more general practice. In his seminal book *Everyday Life in the Modern World* from 1968, the French sociologist Henri Lefebvre proposes the idea that human needs and desires are defined by bureaucracy, and not by the individual. Or put differently, the modern capitalist society programs consumer practices into everyday life, through signs. In a consumer society, products are not only consumed for their practical use but also as signs. Lefebvre gives the example of a car as a "status symbol, it stands for comfort, power, authority and speed."[26] Subsequently, not having a car inevitably implies a lack of, say, authority, while having particular brands of cars inflicts different levels of authority and power (which are frequently negotiated on the highway). In other words, the bureaucracy that created the needs and desires of everyday life constantly assesses and answers how well the consumer "buys into" a lifestyle, with specific hierarchies and codes.

Lefebvre's thinking is not unlike the analyses of Word discussed above, where the tool for word processing simultaneously designs its users to write proficient business-like English, for instance. If the users do not use the right tool (with the best auto-correct function and file format), they not only fail but are also failures; bureaucracy defines not only human desire but human existence, too. Or as Gilles Deleuze asserted, "Machines are social before they are technical."[27] Being human depends on a bureaucracy that encompasses consumer society as such, but in particular manifests in today's platforms for social and cultural software. From quoting text and posting images, to liking, reposting, or disliking them, bureaucracy controls not only the curation and distribution of content but also the very need to produce and consume content. To free oneself from the quotidian, by producing or consuming differently, as Lefebvre proposed, no longer seems a productive escape. Rather, it is somewhat expected and encouraged by a bureaucracy that performs predictive analyses of user-generated data (they simply form new neighborhoods, as described in chapter 1).

The US cultural theorist Ted Striphas has applied Lefebvre's ideas of "controlled consumption" in-depth to contemporary publishing and book trade. Striphas summarizes his analysis into four principles:

1. A cybernetic industrial infrastructure that integrates and handles production, distribution, exchange, and consumption.

2. Programming that closely monitors and tracks consumer behavior along with the effects of marketing, and thereby also reconfigures the relations between consumption and production.
3. Obsolescence that is programmed into the product, limiting its functionality and durability.
4. A disruption and reorganization of everyday life practices.[28]

Even though Striphas uses Lefebvre's ideas to characterize the US book trade, they can also be extended to describe what happens in and around the app interface as well as the metainterface industry.

First, the metainterface industry is characterized by the integration of hardware, software, and distribution network. Not only Amazon's Kindle platform but also platforms from Apple, Google, Microsoft, Sony, and others all feature an ease of use that depends on an insular management system that prevents exchange operations with other, related systems. For example, with an iOS device it is extremely easy to access apps and their content, but it is virtually impossible to download software or media files in other ways than through the App Store or authorized apps. The transfer of files from one platform's information silo to another therefore becomes a convoluted process that quite possibly involves a jailbreaking of the device to remove the software restrictions and permit root access to the system directory. This conveniently ensures control over the publication of content, and the exchange of content that is both monitored and monetized. The physical device is eventually a storefront for the metainterface industry, and if the company behind it should for some reason terminate its activities, the device becomes virtually useless.

As discussed in relation to *RealBeat*, this integration generally prohibits access to the root directories of the devices, and in many ways these platforms are more closed and controlled than ordinary proprietary PC software. The devices function more like advanced media players, or "IT appliances" that hide essential parts of their functionality for the user, as Cory Doctorow argues.[29] They partly lack the universality and open dynamics of the general-purpose computer, and only run authorized programs in protected sandbox environments. As a consequence, software culture becomes limited in developing new, innovative ways of using and understanding the computer, and ultimately in developing new forms of software.

Second, the business model surrounding the integration of hardware, software, and network changes the relations between production and consumption. The products of the metainterface industry are not only consumer products to be bought and sold but also metainterfaces to networked services where users contribute and share information, and simultaneously, where their behaviors are tracked (what they write, what they like, what they share, etc.). This not only inscribes users into an infrastructure that designs them as producers and consumers but also generates large parts of the financial basis of the metainterface industry. It leads to what Shoshana Zuboff, with reference to Google and other "hyperscale" companies, has characterized as "surveillance capitalism." Surveillance capitalism is defined as a "new form of information capitalism [that] aims to predict and modify human behavior as a means to produce revenue and market control."[30] The monitoring of behavior is not only an added bonus for the marketing department but becomes the essential production as well. It is therefore no coincidence that companies like Uber have been portrayed as a big data company that gathers valuable information on user and traffic behavior, and evidently the sharing economy often relies on a simple process of exchange, where access to services is traded for voluntary data capture.[31] Free cloud services and online tools such as Gmail, Google Docs, Google Drive, or Dropbox as well as free social media platforms, free search engines, free media streams, and much more, all come at the same price.[32] In fact, what the user at their human–computer interface experiences as sharing (and as such, something regarded positive) is transformed into the dominating business model: users share, but platforms collect and possess the data in the metainterface (through tracking, datafication, or signal computer interaction, as described in the introduction).[33]

Following Lefebvre's depiction of everyday life in the bureaucracy of controlled consumption, the imperative logic to share and consume in an interface culture reverses the relation between production and consumption: consumption, otherwise perceived as a passive engagement, becomes productive, and conversely, the making and sharing of cultural content, otherwise perceived as an active engagement, becomes integrated into the platform. As the not-for-profit, anticonsumerist, proenvironment organization Adbusters have argued, "Your living room is the factory, [and] the product being manufactured is you."[34] Their claim was broadcast in an "un-commercial" and directed against commercial television, but in a metainterface industry where many services come with little or no content, this seems even truer: user-generated content is often a clear sign that production and sharing comes as a service that is exchanged for the data tracking of consumption patterns (the core business of the platform). Subsequently, the social media platform, cloud service, search engine and streaming service all become important factories of profiling and datafication, although some companies seem more willing to protect users' privacy than others, depending on their reliance on surveillance capitalism as business model.

The reconfiguration of production and consumption also applies to services where users pay for content.[35] Amazon's Kindle e-reader tracks reader behaviors, and the data becomes part of its marketing strategy, or more specifically, prediction of user taste and preferences. The same principle applies to Netflix and many other services that offer cultural content for a revenue, and in some cases this allows the service providers to become actual producers, too (e.g., Netflix not only distributes but also produces television series). The distribution of cultural content is an especially well-suited area to collect data on cultural and aesthetic values, including political preferences and judgments of taste.

Third, the metainterface industry is characterized by the well-known experience of having to endure constant software and hardware updates. The introduction of new file formats, standards, and protocols frequently result in a situation where, for example, newer versions of apps are incompatible with old devices, or the other way around. In other words, updating controls the functionality and durability of cultural computing devices of the metainterface, and this underlines the fact that owners do not independently control their platforms.

Finally, the integration of production, distribution, exchange, and consumption into software and hardware platforms, the tracking of user data that reverses consumption and production, and the programmed obsolescence of software and hardware that limits their functionality and durability all influence cultural practices and everyday life. Collecting, sharing, and creative reconfiguration of cultural content, say—which are all well-known practices with former platforms (books, records, and cassette tapes), and became extensively popular in personal computing's remix and mash-up culture—have become not only illegal but also almost impossible with the platforms of the metainterface. When licensing and streaming replaces ownership, users change their everyday behaviors. Instead of lending a friend a good book or piece of music, users are referred to sharing links to services, for instance. In this way, sharing, which was once seen as a shared human experience of enjoying literature or music, has turned into platform advertisement for Spotify, Amazon, Facebook, and so on (in fact, often several of these platforms at once when a user "shares" a Spotify link on Facebook). In addition, this changes the cultural and democratic institutions around culture. The library, for example, is not only threatened by the digitization of cultural content but more so by the privatization of digital culture, where phenomena such as Amazon's streaming services aim to take over the former role of libraries as collectors of information made accessible to a defined community, yet only to those who pay the license fee. Clearly, the difference between collecting and lending as a public institution, and publishing and licensing as a private business, is a matter of public interest. Even the concepts of reading and literacy are at stake, as discussed below.[36]

The disruption and reorganization of everyday life caused by a new bureaucratic and instrumental control of production, distribution, and consumption is reflected as a tendency within the artworks (following Benjamin's understanding as elaborated in chapter 1). If Fuller's *A Song for Occupation* and Piringer's *nam shub* engaged in an interface culture defined by office work, a similar interest in the metainterface industry that integrates culture in bureaucratic computational platforms can be found today. This tendency, which in many ways follows former critiques of consumer society, often presents a critical attitude toward the new culture that manages to produce ambiguities or subversive alternatives. More important, however, it seeks to take apart the machinery of the new cultural platforms—by reflecting the production system of the metainterface that produces writing and reading, how it "machines" language and composes a sensorium that is at once human and nonhuman, representational and computational, or sign and signal.

Molleindustria's aforementioned *Phone Story* is a good illustration of how the platform is critically reflected within the game, and how this potentially challenges Apple's brand and business model, which builds on the principle of integrating hardware and software (the first principle), and the constant need for updates of software and hardware (the third principle). Such behavior reveals the reconfiguration of cultural curation (the fourth principle)—before exercised by publishers and curators, and now embedded in vague developer guidelines and internal company bureaucracies. Yet the bureaucratic and automated control mechanisms that are involved in large-scale curating and content production, and are programmed into the platforms as surveillance capitalism (the second principle) also reveal other reconfigurations. More specifically, the following examples each present aspects of this in their exploration of a machine language that is not only used by humans (as Word, similarly, is not only a tool for writing but simultaneously produces writing and reading as a designed, instrumental, and industrialized activity):

1. A writing machine that transforms the HCI of word processing into a global machinery of text.
2. A reading machine that reads how the reader reads this global text.
3. A general control of language that has little interest in what is written or read, but great capital interest in the system that produces text and language.
4. A body machine that represents the outcome of this process: a body that is not only designed for consumption to fulfill its existence but also to share and be part of the machinery of the metainterface industry.

Each of these secondary principles describes the metainterface industry that (through, say, surveillance capitalism) causes a reorganization of everyday life, and will in the following be explained through artistic practices that in various ways associate them with the metainterface industry's particular platforms. This by no means implies that the activities of the platforms are restricted to, or have a monopoly on, the particular secondary principles. As the artworks exemplify, the platforms are merely proponents of the principles. Ubermorgen's *The Project Formerly Known as Kindle Forkbomb* addresses Amazon's print press and the production of writing; John Cayley and Daniel Howe's *The Readers Project* focuses on Google along with the changed notion of reading, and their *How It Is in Common Tongue* looks at the capitalizations of language; and JODI's *ZYX* and Erica Scourti's *Body Scan* address the iPhone's and Google's grammatization of the user's body as a new combined and instrumental sensorium.

Amazon Writing: *The Project Formerly Known as Kindle Forkbomb*

Apple's iPhone is obviously an important device for both the metainterface industry and turning cultural computing into an industrialized entity, but it is not the only one. E-readers such as Amazon's Kindle combine old traditions of literature, reading, and book production with the metainterface industry's new traditions of licensing and integrating hardware, software, and distribution networks. Amazon's many activities are good examples of how literature and book production become part of a wider field of textual machineries. The company began as a book retailer in the early days of the World Wide Web, but soon expanded its business. Today, it includes its own media platforms: the Kindle e-book reader and tablets as well as print on demand and publishing services with direct sales through to the Amazon bookstore. In addition, Amazon is a leading provider of cloud services as well as streaming music and video (Prime), retailer of consumer goods, provider of crowdsourced production (Mechanical Turk), and more.

A central element of Amazon's business is to direct its sales strategically. Amazon is renowned for including readers' responses (customer reviews) to the retail experience on its website—a model that has been widely adopted by not only retailers but also other services such as YouTube where users replace the professional critic. Amazon's manufacturing of words has been completely absorbed by complex business structures and machine learning processes that present the newest and most helpful reviews to each customer. This textual business machine might seem advanced compared to the persistency of the traditional printed book, despite the many predictions of its death, but it can also be seen as a continuation of the printing press's technologizing and capitalization of the word.

Walter J. Ong, a historian of print, argues that Johannes Gutenberg's printing press was a groundbreaking technology that led to increased literacy, new modes of reading (silent reading), and renewal of institutions around the study and production of text—such as the library and university. He further asserts that the printing press was the first carrier of industrialized capitalism, leading to a new business around the publishing press, book trade, and newspapers and magazines, with their journalists and critics: "Alphabet letterpress printing, in which each letter was cast on a separate piece of metal, or type, marked a psychological breakthrough of the first order. It embedded the word itself deeply in the manufacturing process and made it into a kind of commodity."[37]

If we compare the printing press with our contemporary textual machinery of the metainterface and a production system like Amazon's, it is clear that the commodification of words not only happens through their reproduction. Instead, the manufacturing of words has been completely absorbed by complex business structures and computational processes. Amazon is an example of how language itself has become the fuel of a new mode of capitalism. The e-reader—as well as social media, search engines, hypertext, programming languages, and so forth—are all illustrations of how the current ecology of text is expanded, and no longer includes just the writing and print of text.

In *The Project Formerly Known as Kindle Forkbomb* (figure 2.4), the artist duo Ubermorgen was inspired by the negative comments on Rebecca Black's *Friday* video—a viral music video on YouTube that received massive attention for its poor quality, with hundreds of thousands of hateful comments and numerous popular parodies. In 2012, Ubermorgen collaborated with Luc Gross and Bernhard Bauch to build an Internet robot that could automatically generate books on the basis of YouTube comments on videos, and then upload them as e-books to Amazon's Kindle bookstore—producing a whole literary ecology including crowds, authors, books, titles, accounts, pricing, and a defense system against erasure. In this way, the system emulates Amazon's own platform by making a parasitic system inside of it, and it functions as a demonstration of the mechanisms of today's production of text.[38] It might even be seen as a perverted version of the ideal postprint capitalism that delivers lots of cheap goods in the form of e-books to fill the endless shelves at Amazon's electronic warehouses. Just like the books by Philip M. Parker (who is currently the author of 127,174 books on Amazon), produced from databases and Internet searches, *The Project Formerly Known as Kindle Forkbomb* demonstrates a new fully automated book production, which can fill the

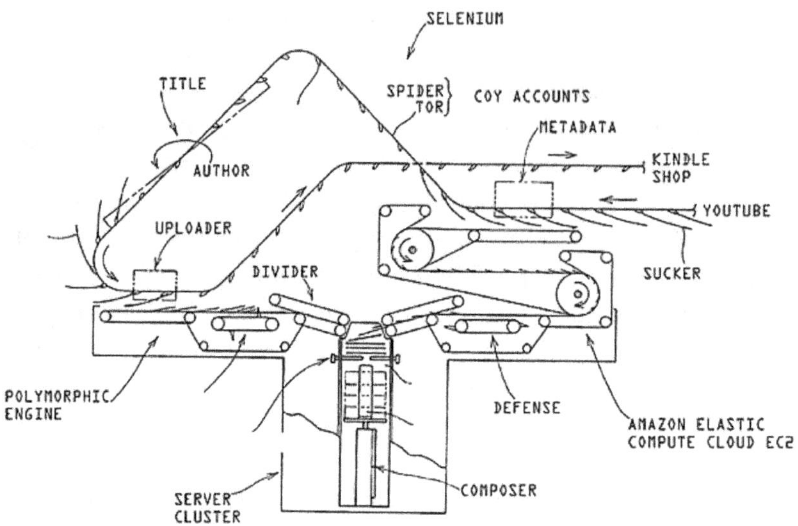

Figure 2.4
The Project Formerly Known as Kindle Forkbomb (2011–2013) by Ubermorgen. A hybrid printing press system that automatically generates e-books on Amazon's Kindle bookstore. The texts are created on the basis of YouTube comments on videos. Courtesy of the artist and Carroll/Fletcher London. Diagram by Ubermorgen.

still-growing storage space on the consumers' Kindles. While Parker produces titles like *The Official Patient's Sourcebook on Narcolepsy* with infinite boring reports, though, *The Project Formerly Known as Kindle Forkbomb* produces a new kind of staged drama.

The books produced by the system appear as dramas with actors performing (or rather producing) social media. In the book *You Funny Get Car* by "Nrlnick Kencals," we meet thirteen characters discussing a video in which the teenage pop idol Justin Bieber appears anesthetized in the back seat of a car.[39] As often happens on YouTube, the comments are hateful (the comments to Rebecca Black's *Friday* video is another example of this), and sexually harassing toward Bieber ("when you press eight, thats when biber has a dick rammed up his ass !"). On a closer look, however, the dialogue is much more complex. Apparently Bieber's video is part of another phenomenon, *Funny or Die*. Funnyordie.com, a website founded by Will Ferrell and Adam McKay, which allows famous people to upload their own funny videos. The videos are rated by the users: the funny ones stay, and the others are archived in the "crypt." To further complicate the dialogue in the drama, the characters discuss the sophistication of Bieber's prank, hinting that he is not acting anesthetized but instead reenacting another YouTube hit video in which a young boy is on his way home from the dentist.[40]

In other words, reading the text is a complex affair. Dramas are renowned for their structuring plot, and one's ability to read reflects the ability to detect this layer of the text. In Ubermorgen's drama, though, it is almost impossible to find the structuring layer. Behind every veiled sign there is another veil of implicit references to network culture. *The Project Formerly Known as Kindle Forkbomb* reflects—almost as a mirror or kind of realism—the production of language and text in the metainterface industry's popular culture. What makes the text in Ubermorgen's e-book is neither the video of Bieber on YouTube nor its original context at funnyordie.com, nor even the video of the anesthetized boy on his way home from the dentist. It is the relations between these texts and the different levels of experiences that they reflect. Subsequently, the ideal reading experience does not find a structuring plot but rather an ecology of texts and relations that produces meaning. These relations must be seen as a textual and semiotic manifestation of a human reality, yet whereas the production of meaning traditionally is understood as related to a human condition (language as a human activity), *You Funny Get Car* reveals relations to a text machinery: if symbol activity was once a human affair, Ubermorgen demonstrates how it is embedded in a machinery that commodifies words.

In Ubermorgen's work, the production of text is both the effect of human and machine agencies. As a fractured expression of a new kind of text production, which mimics the text machinery of the networked computer, the work reveals an inner tendency. The project works as a "forkbomb" in the textual machinery of the metainterface industry. In computing, a forkbomb is a denial-of-service attack, meaning it is a process that continually replicates itself inside the system, draining it of its resources like a parasite and ultimately causing the system to crash. These automations constantly evolve around the semio-capitalist text machinery that produces our reality, but Ubermorgen automatizes the networked processes that produce text today and turns text against its own textual machine, as a forkbomb.

Ubermorgen's forkbomb can be seen as a contemporary form of realism through the way it explores how current expressions of text (e.g., comments on YouTube) are intrinsically linked to the production of reality embedded in the machinery (e.g., the processes of harvesting text and embedding it in new contexts), and in this way it connects to a literary history of realism going back to classic French realism of the 1800s. The Marxist literary critic Georg Lukács has pointed out how literary realism explores the ways that technological developments affect culture and how art can respond to this. He did this in a reading of the French nineteenth-century realist Honoré de Balzac's *Lost Illusions*, which he read as a novel about how literature became commodified in a capitalistic production system, and how this lead to a capitalization of ideology and thinking.[41] To Lukács, Balzac's novel is an example of a conscious exploration of a new discourse economy, and it shows how material changes influence the formal conditions of the artwork.[42]

The Project Formerly Known as Kindle Forkbomb might not look like typical realistic literature in the manner of French literary realism, but Lukács's reading of Balzac nevertheless makes it possible to discuss it as realism. It reflects how current expressions of text (e.g., comments on YouTube) are intrinsically linked to the production of a reality embedded in a language machinery (e.g., the processes of harvesting text and embedding it in new contexts). But its realism is also different from Balzac's. Though partly a work of popular culture, *You Funny Get Car* is not a popular work. The video comments are of limited interest outside their original context, and the characters appear as one blurry, homophobic, judgmental crowd. The drama is present, but appears as if it was painted with one color only. Instead of selling critique as literary consumption, it shifts focus and sells it as literary production—a literature machine. It turns the bureaucratization of culture (e.g., in Amazon's replacement of expert critics with user recommendations, algorithmic profiling, and recommendations) into the product of its literary machine and text of its dramas, thereby also showing how this text is constituted in networks and contains its downsides, as when judgments of taste are replaced by racist and homophobic flaming. In this sense, acknowledging that it is not only human language that produces our reality but also the ecology of the networked computer's text machinery, the literary project is to develop a poiesis of the textual machine, to produce a language reality with the machine. The forkbomb at once mimics and mocks the textual machine of the metainterface industry; it produces an automated and endless long tail of more or less unreadable books, echoing Philip M. Parker's extensive book production—books that probably nobody reads, writes, or even discovers, since they are strategically kept secret from Amazon in order to keep the forkbomb running inside the system as a hidden parasite. It explores the tendency of Amazon's printing press through the parasitic production system of the Kindle forkbomb, and in this way, like Balzac's novels, demonstrates how the functional basis of the metainterface's production system and the cultural and literary, artistic superstructure of writing are related.

As much as Amazon is a post-Gutenberg printing press that prints and distributes books on demand, it is also a system of reading. If Amazon is renowned for its inclusion of customer reviews in the retail experience, the store is renowned, too, for using behavioral data along with peoples' taste and preferences to generate consumer patterns and predictions. Today, Amazon even tracks patterns of reading. The users' reading patterns are handled through Amazon's Whispernet, a cloud service that connects to the Amazon platform and stores reading data while the user is reading, including what, when, and where the user reads, and which notes and underlinings are made. This kind of data capture is no longer only used to offer products on its website, but is increasingly integrated into publishing and distribution as prediction models for reader interests and purchases. So not only does Amazon provide a "one-click" shopping experience and recommend products to the users, it also delivers fast and exemplifies how machine learning can be implemented in business, but it is produced through monitoring reading behavior on the web and Kindle.

Google Reading: *The Readers Project*

Text production has always been a central part of the development of the World Wide Web and its infrastructures. This text production not only includes hypertext in its purest sense but is a complex conglomerate of references and citations spun together by a machinery of writing as well; YouTube comments and reference to other social media may suddenly appear in e-books, as Ubermorgen demonstrates. A central factor in the text machinery is Google. Google is naturally renowned for its search engine, web browser, Android operating system, and many cloud-based web services that include maps, e-mail, word processing, file sharing, and more. It also owns YouTube, market its own hardware platforms, and develops designs for imaginary futures, such as its extensive development of self-driving cars. As a large-scale service provider on the Web, advertisement agency, and information technology developer, it is a key player in a metainterface industry. Despite the scale and variety of Google's activities, its control of production, distribution, and consumption of the reading of text in particular seems to be a core activity at a scale that far exceeds Amazon's. For instance, the reading of a Google search result is a controlled experience based on a complex reading of how and what people read that Google captures from its various services.

This double-sided reading that characterizes the metainterface—where machines read how humans read—is not new since interaction has always conditioned what computers register and how they react to users' behavior at the interface, but it has spread and been increasingly intensified with the metainterface industry and surveillance capitalism. We increasingly read texts that—like *You Funny Get Car* or a search result—are rewritten by algorithms programmed to mimic and manage our reading. Or put differently, the texts we read integrate a large body of text, and the scripts that control this integration are (in more or less sophisticated ways) based on scripts that monitor reading behaviors. The conditions of reading are in this way significantly reconfigured by the metainterface industry. In his book *Scripting Reading Motions*, the Portuguese digital literary theorist Manuel Portela analyses and discusses digital literature and literary interfaces that explore how human reading motions are being scripted into the interfaces. These self-reflexive interfaces mimic how algorithmic reading simulates, tracks, and rewrites human reading, and demonstrate a battleground between human and algorithmic reading where readers are confronted with the effects as well as politics of the changed conditions of reading. As Portela describes it, "The material instantiation of text, and our perceptual consciousness of it, respond to and reflect our own haptic and visual interaction with the interface."[43] One of his examples is *The Readers Project* created by the artists, writers, and theorists John Cayley and Daniel Howe.[44]

The Readers Project (figure 2.5) consists of a series of ongoing experiments, installations, and performances that relate to reading. These experiments are based on literary

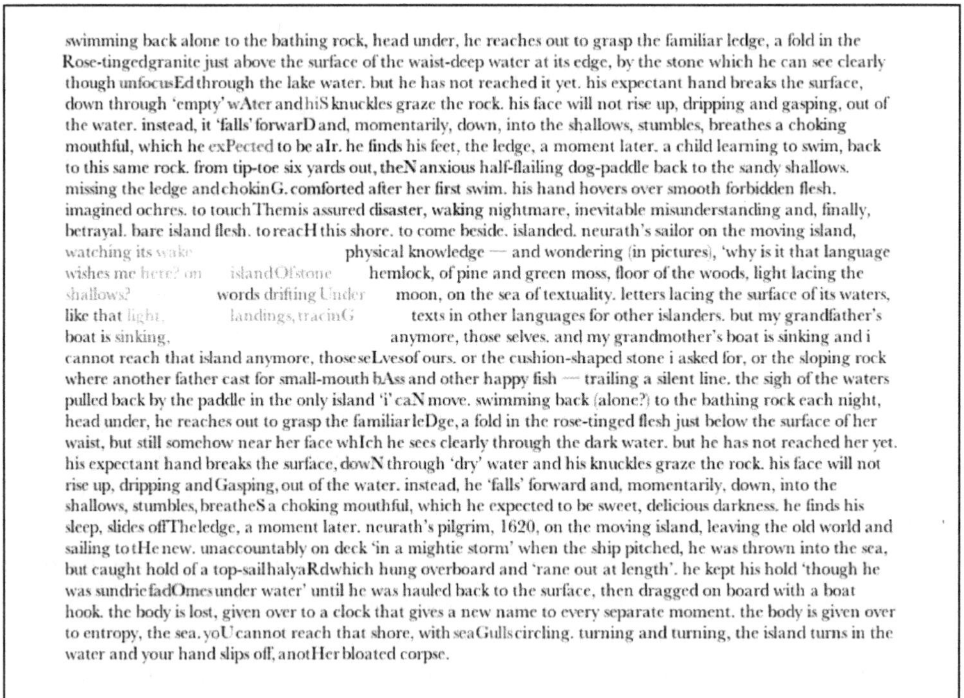

swimming back alone to the bathing rock, head under, he reaches out to grasp the familiar ledge, a fold in the
Rose-tingedgranite just above the surface of the waist-deep water at its edge, by the stone which he can see clearly
though unfocusEd through the lake water. but he has not reached it yet. his expectant hand breaks the surface,
down through 'empty' wAter and hiS knuckles graze the rock. his face will not rise up, dripping and gasping, out of
the water. instead, it 'falls' forwarD and, momentarily, down, into the shallows, stumbles, breathes a choking
mouthful, which he exPected to be aIr. he finds his feet, the ledge, a moment later. a child learning to swim, back
to this same rock. from tip-toe six yards out, theN anxious half-flailing dog-paddle back to the sandy shallows.
missing the ledge and chokinG. comforted after her first swim. his hand hovers over smooth forbidden flesh.
imagined ochres. to touch Them is assured disaster, waking nightmare, inevitable misunderstanding and, finally,
betrayal. bare island flesh. to reacH this shore. to come beside. islanded. neurath's sailor on the moving island,
watching its wake physical knowledge — and wondering (in pictures), 'why is it that language
wishes me here? on islandOfstone hemlock, of pine and green moss, floor of the woods, light lacing the
shallows? words drifting Under moon, on the sea of textuality. letters lacing the surface of its waters,
like that light, landings, tracinG texts in other languages for other islanders. but my grandfather's
boat is sinking, anymore, those selves. and my grandmother's boat is sinking and i
cannot reach that island anymore, those seLves of ours. or the cushion-shaped stone i asked for, or the sloping rock
where another father cast for small-mouth bAss and other happy fish — trailing a silent line. the sigh of the waters
pulled back by the paddle in the only island 'i' caN move. swimming back (alone?) to the bathing rock each night,
head under, he reaches out to grasp the familiar leDge, a fold in the rose-tinged flesh just below the surface of her
waist, but still somehow near her face whIch he sees clearly through the dark water. but he has not reached her yet.
his expectant hand breaks the surface, dowN through 'dry' water and his knuckles graze the rock. his face will not
rise up, dripping and Gasping, out of the water. instead, he 'falls' forward and, momentarily, down, into the
shallows, stumbles, breatheS a choking mouthful, which he expected to be sweet, delicious darkness. he finds his
sleep, slides off The ledge, a moment later. neurath's pilgrim, 1620, on the moving island, leaving the old world and
sailing to tHe new. unaccountably on deck 'in a mightie storm' when the ship pitched, he was thrown into the sea,
but caught hold of a top-sail halyaRd which hung overboard and 'rane out at length'. he kept his hold 'though he
was sundrie fadOmes under water' until he was hauled back to the surface, then dragged on board with a boat
hook. the body is lost, given over to a clock that gives a new name to every separate moment. the body is given over
to entropy, the sea. yoU cannot reach that shore, with seaGulls circling. turning and turning, the island turns in the
water and your hand slips off, anotHer bloated corpse.

Figure 2.5

The Readers Project (2009–) by John Cayley and Daniel Howe. A series of installations and software that makes programmed reading visible and readable to the human reader, including a "perigram reader" that is defined by a typographic neighborhood and a "mesostic reader" that follows specific letters that form words. Courtesy of John Cayley and Daniel Howe. Screen capture from software.

software, or programmed readers, that read texts, rewrite the texts, and present them to human readers, thereby making their reading visible and readable for the human reader. In other words, their literary interfaces visualize the programmed readers' reading. The programmed readers include Simple, Perigram, Spawning, Mesostic, and Grammatical Lookahead Readers, and they can be downloaded as Java Archives. Their reading patterns are inspired by cognitive studies of human reading, and range from something close to standard Western human reading (from left to right and top to bottom in the Simple Readers), to reading across what Cayley and Howe define as the typographic neighborhood and page (Perigram Reader), to readers looking for specific letters in order to form words (Mesostic Reader), and readers following the grammatical structure of the text and finding alternative words to fit this (Grammatical Lookahead Reader).[45] Consequently, the different vectors of reading create routes through the text based on algorithmic rules, typographic neighborhoods, or grammatical and semantic structures. As Portela writes, "The physical, cognitive, and social dimensions of the process of reading are algorithmically simulated in a way that invites readers to see their own act of reading in action."[46]

In some of the interfaces, the human readers can only read the texts through the programmed readers' reading and rewriting of the texts; in others, the programmed readers' routes are highlighted and obviously influence the human reading. The human readers thereby not only become conscious of their own reading process (including the grammars, habits, and materials governing it) but also the algorithmic readers' grammars and (re)writing of the text. This double reading-writing ultimately becomes visible through the way that the human reading is blocked by the programmed readers' performance. Since it is only partly visible and gets disturbed by the programmed readers' zigzagging through the text, it is often difficult to read the original text. Instead, the human reader reads how the text is generated and rewritten by the programmed readers. The grammars of literacy literally—displayed in the interface—clash with the grammars of the reading algorithms. HCI becomes rewritten by signal computer interaction in the metainterface.

In this way, the mainly invisible algorithmic paratextual framing by algorithmic agents (such as Google's spell and grammar checking, indexing and search bots, contextual advertising, other people's popular highlights, etc.) becomes visible. Cayley and Howe specifically mention the writing of their own text in the Google Docs interface as an example of this paratextual framing by algorithmic agents.[47] In this way, the project demonstrates and radicalizes the way paratextual framing agents influence our reading. As Portela argues, the project is "scripting the act of reading. ... Since these readings of the machine are offered as writing to human readers, the writing of reading and the reading of writing contained in *The Readers Project* turn readers into metareaders who are forced to read their own act of reading the program reading."[48] In the programmed readers' interfaces and real time, the human reader is able to read the comprehensible and incomprehensible mechanisms that complicate the textual dynamics, and see how reading and the readable is not only governed by the reader but also increasingly by programmed dynamic agents creating alternative connections and neighborhoods. The human reader thereby sees that they are not the only reader, and that the readable is cowritten as well as governed by networked software and corporations in the cloud that are mimicking, tracking, and rewriting the reading. The human reader ultimately metareads, and realizes that his or her reading is enmeshed in a networked cybertext where reading is tracked and used to generate writing in an endless data loop that we also know from social media, yet rarely are able to read directly.

Through the production of reading, as an inner tendency, the work reflects the changed conditions of reading. This is revealed as cracks and tensions within the work where passive reading becomes an active form of rewriting. In the metainterface industry, reading (and consuming, more generally) becomes a production, but in *The Readers Project*, the human reader experiences how the text becomes controlled, too, and how this challenges his or her reading. *The Readers Project* lets us read how our reading becomes productive as rewriting, and how this production becomes part of the text and textual business of big software companies such as Google. The reader is able to see, explore, and read the bureaucratization and instrumentalization of reading.

Google Language: *How It Is in Common Tongues*

The changed conditions of reading along with the encounter between human and machine reading is also a political battle that extends a former copyright conflict of the World Wide Web into the metainterface industry. As part of *The Readers Project*, Cayley and Howe have produced the artists' book *How It Is in Common Tongues* (figure 2.6).[49] The book is generated through a script that uses Google to search for the longest strings of text that match Samuel Beckett's *How It Is*, while avoiding text actually written by Beckett. The result is a book with the same text as Beckett's original, but quoted from other texts written by other people, and not least an enormous amount of footnotes with URL references to these texts. In other words, the text is cut up in short sequences of typically two to five words caused by the longest portion of Beckett's text that could be found in other texts. Interestingly, this is equal to the way humans read texts in chunks of words. This was also emphasized by Cayley and Howe in the exhibition his[2d4] aid sits a[2d5] little aloof he[2d6] announces brief[2d7] movements of the lower face the[2d8] aid enters it[2d9] in his ledger

my hand[2da] won't come words won't come[2db] no word not even[2dc] soundless I'm in[2dd] need of a word of[2de] my hand dire[2df] need I can't[2e0] they won't that too[2e1]

deterioration of the sense of humour[2e2] fewer tears too[2e3] that too they are failing[2e4] too and there another[2e5] image yet another[2e6] a boy sitting on a bed[2e7] in the dark or a small[2e8] old man I can't see[2e9] with his head be it[2ea] young or be it old[2eb] his head in his hands I[2ec] appropriate that heart[2ed]

question am I happy in the[2ee] present still such[2ef] ancient things a little[2f0] happy on and off[2f1] part one before[2f2] Pim brief[2f3] void and barely[2f4] audible no no I[2f5] would feel it and[2f6] brief[2f7] apos-

[2d4]www.niemanlab.org/author/kdoctor/ (id. 548000)
[2d5]www.fold3.com/document/84474314/ (id. 54) [2d6]casanctuary.org/freedom/ (id. 187)
[2d7]www.shakermakerpr.com/post/25379254492 (id. 26400) [2d8]www.ask.com/Lower Face Lift (id. 1) [2d9]forums.soompi.com/discussion/136622/bitexme/p25 (id. 5)
[2da]issuu.com/lsmedia/docs/hasc_low_res (id. 1) [2db]catmp3.com/Sherald.html (id. 10)
[2dc]en.wikiquote.org/wiki/The_Dark_Tower_(series) (id. 664000)
[2dd]www.swjfj.com/archives/120 (id. 9) [2de]www.jstor.org/stable/1487868 (id. 5570000) [2df]fr-fr.facebook.com/isaure.palfroy (id. 4)
[2e0]www.youtube.com/all_comments?v=gm8TTTthKzQ&page=1 (id. 55900)
[2e1]wtmcclendon.wordpress.com/page/11/ (id. 6) [2e2]wikiality.wikia.com/Zeezumrazzumprofen (id. 1) [2e3]neurotalk.psychcentral.com/thread101407-32.html (id. 366)
[2e4]letters.mobile.salon.com/opinion/feature/2009/06/08/wingnut/permalink/99ff69f384498160c4b223b75de38e99.html (id. 1) [2e5]web.stagram.com/n/meggmack/ (id. 1430000)
[2e6]waxy.org/links/archive/2004/01/ (id. 2160) [2e7]www.greglast.com/?page_id=156 (id. 222000) [2e8]chasethetruth.blogspot.com/ (id. 2) [2e9]thebarefoot.wordpress.com/2012/03/ (id. 18200) [2ea]scoopssn.blogspot.com/ (id. 3) [2eb]airmax2010good.blogspot.com/ (id. 36) [2ec]i-gave-up.com/?s= (id. 932000) [2ed]www.ocezine.com/page/page/list (id. 63)
[2ee]www.amichopine.com/blog/?p=140 (id. 10) [2ef]www.mtb-bg.com/forum/viewtopic.php?f=6&t=310 (id. 10)
[2f0]s-clothing.blogspot.com/2010/04/70s-clothing.html (id. 6) [2f1]www.mgoblue.com/ (id. 46600) [2f2]nongcavenderphotography.com/?p=3066 (id. 760000)
[2f3]www.microform.co.uk/guides/R97280.pdf (id. 5)
[2f4]andywhitman.blogspot.com/2009_11_01_archive.html (id. 7) [2f5]www.amazon.com/ (id. 2)
[2f6]www.ziglar.com/newsletter/?p=1595 (id. 270000) [2f7]brief.mozdev.org/ (id. 398000000)

[30]

Figure 2.6
How It Is in Common Tongues (2012) by John Cayley and Daniel Howe. A book generated by a script that uses Google to search for the longest strings of text that matches Samuel Beckett's *How it is.* In other words, Beckett's text is regenerated from quotes that are not written by himself, but are "common tongue." Courtesy of John Cayley and Daniel Howe. Screen capture of e-book.

Common Tongues, where the book was displayed next to an interface showing a perceptual reader reading through parts of *How It Is*.

Yet Cayley and Howe furthermore used the sequences of words to search online for neighboring texts not associated with Beckett.[50] In this way, they not only displayed a parallel between algorithmic and human reading (or signal–computer interaction and HCI in the metainterface) but also demonstrated how Beckett's *How It Is* relates to a network of free text and language—a "common tongue" that exists outside Beckett's ownership and copyright. The issues of copyright and ownership of language are furthermore reflected in the licensing of the book as "Copyleft, *hors de commerce*": it is an artwork "generated from orthothetic links into publically accessible records from the great inalienable linguistic commons," as Cayley and Howe explain.[51] They explicitly describe their work as conceptual algorithmic art built on an original concept, and not as a work of literature, which would violate Beckett's copyright.

Put differently, Cayley and Howe point to how even copyrighted work speaks in common tongues, and how contextual *work* is always composed by and related to a common *text* (to reference Roland Barthes's famous essay *From Work to Text*).[52] In this way, *How It Is in Common Tongues* demonstrates that words are not only part of the arrangement of sentences, and thereby the copyrighted work, but also of language as a network of common text. To be more precise, it points out how there is a difference between language as a common abstract and paradigmatic system where words can be changed and reused, and the particular syntagmatic instantiations of language that forms the positions of words in a sentence. As emphasized by the French founder of modern linguistics and semiotics, Ferdinand de Saussure, there would be no meaningful speaking (*parole*) without a language (*langue*), but in the metainterface industry, language—and not only its particular instantiations—is increasingly capitalized.[53]

If *How It Is in Common Tongues* illustrates that Beckett does not control his work as *text*, Cayley and Howe also show that Google currently controls the whole paradigmatic network of text. Google's business concept is, in other words, closely related to managing the paradigmatic network of language. This does not mean that Google and the metainterface industry control meaning in a totalitarian or indoctrinating way; rather, it points to how Google manages, tracks, and indexes the paradigmatic system, and thus capitalizes language. In her discussion of Google, Zuboff characterizes this change as follows: "What matters is quantity not quality. Another way of saying this is that Google is 'formally indifferent' to what its users say or do, as long as they say it and do it in ways that Google can capture and convert into data. ... Individual users' meanings are of no interest to Google or other firms in this chain."[54] Google (and other similar big software companies like Facebook and Amazon) is not interested in *what* we say but rather *how* we use language. Mapping language (*langue*) is its business, which is obviously efficient for forming neighborhoods based on how language is used and how it is biased, but (as it will be discussed below in relation to Scourti) these biases also gets reproduced by Google's (and others') language machine, such as when it suggests searches or search results, or when Facebook's algorithms produces newsfeeds according to its user profiles.[55]

Conceptualizing, mapping, and organizing the network of texts through techniques of referencing and indexing is not a new development in literary culture. References, associations, and intertextuality add new dimensions to a text, and attach the text to greater networks of literature and language. Vannevar Bush famously argued that knowledge is produced through association, and described his conceptual hypertext system, the Memex, as a support mechanism for associative reading and writing.[56] When reading *How It Is in Common Tongues*, one is confronted with the multidimensionality of Beckett's text, how an endless paradigm of other texts is folded into Beckett's, and his into theirs. As a reader, one gradually unfolds one's reading of the text as a paradigmatic networked text rather than as a sequential and syntagmatic work, and Howe and Cayley thus demonstrate that Google does something similar through its indexing of the World Wide Web. Beckett's original text has no traditional punctuation, but is in Howe and Cayley's version punctuated by footnotes with URL references that act as fractures, illuminating how Beckett's text is part of a common tongue. It shows how Beckett's text is a network that feeds on and feeds into other texts. This unfolding of language and limitless networking of *How It Is* is a poetic and potentially sublime experience, setting the work free as text, equal to the way that Cayley describes searching as "literary sublime" and "an encounter with overwhelming quantities of language, arguably beautiful."[57] *How It Is in Common Tongues* makes the reader glimpse this sublime network of tongues, language, and people writing. Besides their own reading, the reader sees the global reading of the common tongues. Yet the reader also experiences how this network has become technologized, bureaucratized, and instrumentalized.

Both Beckett's *How It Is* and Cayley and Howe's *How It Is in Common Tongues* simultaneously point to the interruption of language along with the impossibility of expression: "my hand[2da] won't come words won't come[2db] no word not even[2dc] soundless I'm in[2dd] need of a word of[2de] my hand dire[2df] need I can't[2e0] they won't that too[2e1]"[58] To Beckett's narrator, creative expression is problematic, but the footnotes generated by Cayley and Howe's script add another layer to this. As observed by Ong in relation to the printing press, the "technologizing of the word," including the construction of indexes, references, and notes, shows the "disengagement of words from discourse."[59] The deployment of algorithmic agents and network services to language continues this indexing further, and develops a language that is less discursive and more controllable. Ong also remarks that already the printing press and typography "made the word into a commodity."[60] This commodification is unquestionably intensified with agents and networked services such as the ones actuated by Google. *How It Is in Common Tongues* demonstrates how Google's commodification adds an interface between the writing (Saussure's *parole*) and the language (Saussure's *langue*), or the work and the text. It thereby interrupts the individual reading through the networked reading, and ultimately changes the character of language.

As Howe and Cayley argue, the "algorithmic, compositional, and configurative agents of big software's network services" have changed the nature of reading and writing "far more deeply than even the development of hypertext in the 1990s."[61] *How It Is in Common Tongues* demonstrates how hypertext theorists' (such as Ted Nelson's) dreams of a sublime connected language system has led to the Google's technologizing of the word. This process is not only a continuation of former control mechanisms from the printing press but also continues with the programmed machine reading and rewriting of our language. *How It Is in Common Tongues* illustrates how Google manages this network of language by tracking, saving, and licensing it back to us through their terms of use. With programmed readers that have developed into an underlying network, Google is hypertext's shadow: a metainterface industry that manages the unruly, rhizomatic network of hypertext and turns it into a thriving business; a capitalistic network machine reading our reading and writing, and furthermore this machine is not openly readable in return.

In a poem and article related to *The Readers Project*, Cayley offers the critique that he is not allowed to make algorithmic agents search or read Google in the same way that Google makes algorithmic agents read his texts.[62] Not unlike Apple's control of the iOS platform, and general strategies of the metainterface industry, Google prevents this by enforcing strict terms of use and licenses supported by technical constraints. In this way, *How It Is in Common Tongues* is not only a potential violation of Beckett's old-fashioned copyright but also a potential violation of Google's terms of use and its amassed text. As Cayley writes,

> What if I, a good human, write (that is create or compose) a program that acts like a bad robot for good reasons, for aesthetic, culturally critical reasons, or simply to recapture and reclaim some of that superb big data that lies on the other side of the mouth-threshold where the powerful indexes dwell? Well, if I do that, it's pretty bad, and it's against most terms of use. Big software can, it seems—via innovation, hyper-historical momentum, and force majeure—deploy whatever robots it wishes—to index the web pages that humans have written or to police human access to its services. … But any robot that you or I build and that interacts with these services is "bad" by default, guilty until proven innocent.[63]

Cayley ends his poem "Pentameters" by stating that "to make art on terms? Impossible. / For the sake of art and for the sake / Of every cultural institution and their futures / We must find a way to refuse such / *Terms of use.*"[64] Nevertheless, Cayley and Howe do not refuse the terms—neither of language nor technical protocols, algorithmic agents and big software—but use them in order to consider the conditions of language. Through its own writing, as an inner tendency, *The Readers Project* mirrors the algorithmic agents and networked text that condition reading and language in the metainterface.

iPhone Bodies: *ZYX* and *Body Scan*

As both Molleindustria's *Phone Story* and Cayley and Howe's *Readers Project* demonstrate, it can be difficult to expose and critique the cultural as well as political reality of the metainterface industry from within its own platforms. Attempts are consistent, however. Another exemption that confirms this rule is the app *ZYX*, developed by Joan Hemskerk and Dirk Paesmans, also known by their artist group name, JODI, which adds an embodied perspective to the critique.[65]

JODI is considered a pioneer in net art as well as for its experiments with video games, maps, social media, and other everyday software interfaces. Often JODI's work explores the codes, platforms, and technologies that are normally hidden behind the metaphors, such as the page metaphor of the World Wide Web or immersive three-dimensional environments of first-person games. JODI's deconstruction of the interface is never trivial or self-contained. Rather than being modernistic or media centric, it looks at the tendencies hidden within the artwork's own production, and in a precise and concrete way, points to the culture, politics, and technologies surrounding interfaces. This kind of deconstruction of the interface is also applied in its app from 2012, *ZYX*.

ZYX has two user interfaces: the smartphone app and a corresponding website (figure 2.7). The app can be characterized as a performance tool asking the user to, without further explanation, do repetitive tasks in a series—for instance, to spin around, jump, or blow into the microphone. When the tasks are completed, an alarm goes off and images taken during the performance are shown on the screen. In this way, the app demonstrates how, with a reference to Agre, smartphones imply particular "grammars of actions."[66] By overexposing the normally functional and unnoticed physical relationship between interface and user, the app shows how the interface inscribes the movements and positions of its users, and how apps constantly respond to these movements by measuring and counting them. Moreover, as demonstrated by videos

a.

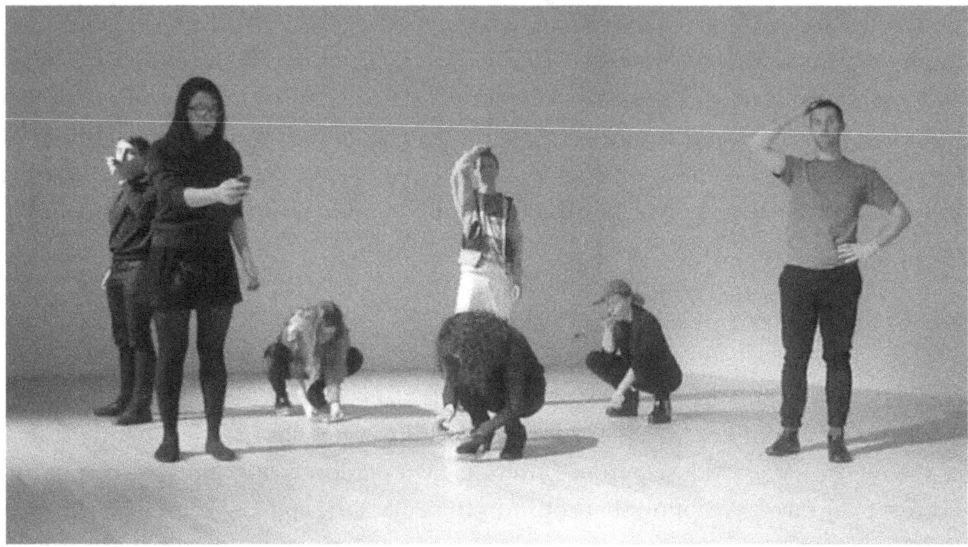

b.

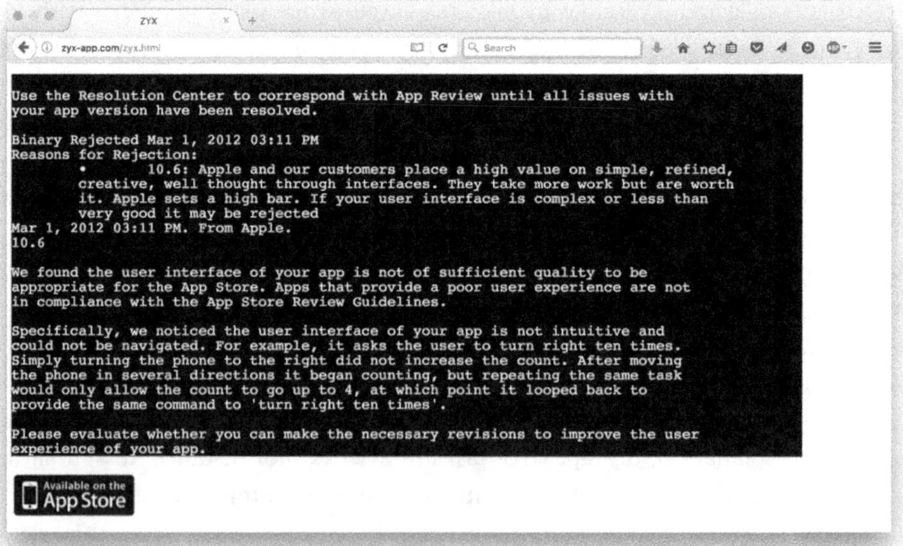

Figure 2.7

ZYX (2012) by Jodi. (a) A smart phone app that asks the user to do repetitive tasks in a series (spin around, jump, etc.). The app over-exposes a normally functional and unnoticed physical relationship between interface and user. Photo by Jodi. (b) *ZYX*'s corresponding website documents the apps hidden interface, including the correspondence with Apple. Screen capture of project website. Courtesy of Jodi.

on the website, interfacing with the app becomes a performance. Without question, the behaviors of the users, who navigate the virtual realm of the app, look bizarre to bystanders. They "see the user as performing a strange dance," as JODI phrases it.[67] As JODI also argues, however, this strange dance has become an everyday phenomenon performed by millions of smartphone users who are desperately trying to find a signal by holding their phone into the air, taking panorama shots, and so on.[68] These strange behaviors often occur because the apps take advantage of the built-in motion coprocessor, which tracks user movement through and collects sensor data from the integrated accelerometer, gyroscope, compass, and GPS.

Already in 1934, the French sociologist Marcel Mauss observed that the girls in Paris had started walking the same way as in New York thanks to Hollywood cinema: "The positions of the arms and hands while walking form a social idiosyncracy, they are not simply a product of some purely individual, almost completely psychical arrangements and mechanisms."[69] With smartphones, lots of apps (maps, games, fitness trackers, and other apps in the quantified-self genre) track and control user movement, and the "social idiosyncracy" is increasingly integrated into the metainterface industry. *ZYX* might be seen as a critical deconstruction of how such apps interface with users—a way of exploring their strange dance and how the interface controls their movement.

Whereas the app deconstructs the interface of the iPhone, the website continues this deconstruction with a view into the normally hidden aspects of the interface. This includes not only descriptions and illustrations of the app but also bits of its program code that exemplify how the app has addressed the iPhone interfaces (e.g., the sensors, camera, and microphone). Interestingly, it reports on the interaction between JODI and Apple's App Store as well. *ZYX* was first rejected because of "poor user experience" and only later approved—this time without further explanation.[70] Both the program code and correspondence reveal the existence of other interfaces to the metainterface industry than the app itself. These interfaces are usually not visible to users, but JODI exposes them as an integral part of the industry's production chain, including how the app needs to be registered and accepted by the App Store. In this way, they demonstrate the bureaucratization of interface culture that follows with the metainterface and its platforms.

Scourti, a Greek artist based in Britain, also explores how users may express themselves and write with interfaces even as they are simultaneously written by the interfaces. In this sense, her work is an examination of a process of gendering, individuation and subjectivization in the metainterface industry. Among other things, she has published the "ghostwritten memoir" *The Outage* "based on her digital footprints" and "told by J. A. Harrington."[71] With a feminist perspective, she often focuses on the body

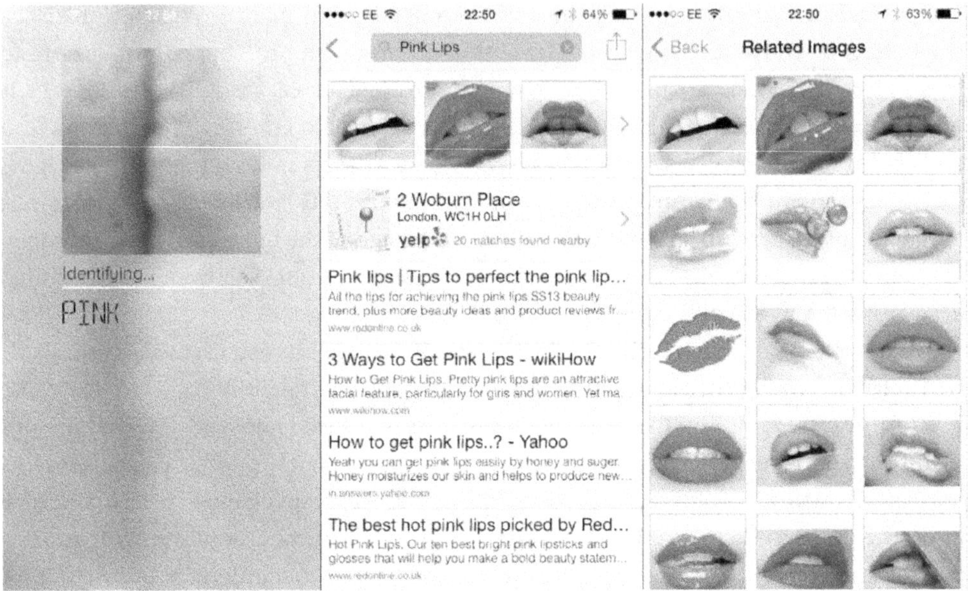

Figure 2.8
Body Scan (2014) by Erica Scourti. In the video, Scourti uses the app CamFind to photograph parts of her body and search for similar images on the World Wide Web. In the corresponding audio track, Scourti reads from the search results. Courtesy of Erica Scourti. Screen captures from video.

along with the ways it becomes profiled, programmed, and gendered through, for example, Google and smartphones, or how Mauss's "social idiosyncracy" is integrated into the metainterface industry. In her video work *Body Scan* (figure 2.8), she uses the app CamFind to photograph parts of her body with an iPhone and search for similar images on the World Wide Web. The video work displays the photographed images and the images returned from the search. In the background, Scourti reads from the search results: "Identifying human stomach. A hollow muscular organ. Forms gastric acid. Is it cancer? And, how to survive another human. Human belly. Human belly button. Why do I have a belly button?"[72]

In her thesis "The Female Fool: Subversive Approaches to the Techno-Social Mediation of Femininity," she analyses the work of classic feminist performance artists from 1960s until now (Valie Export, Martha Rosler, Pipilotti Risti, etc.), and the role of media technologies in their representation—how their "embodied subjectivity [is] inseparable from technology."[73] She continues with a quote by Katherine Hayles that almost echoes Mauss: "When changes in incorporating practices take place they are often linked with new technologies that affect how people use their bodies and experience space and time."[74] Naturally this technological integration of the body is not just something experienced by the perceiving body but also "a reality of life, which profoundly affects the ways female bodies are mediated and experienced."[75]

Similar to JODI's *ZYX*, *Body Scan* demonstrates how the interface has become mobile and intimate, and almost a part of the user's body—defining its very mobility and appearance in space. The works indicate how the bodily effect of media technology is not limited to the perceiving body but instead is part of a larger media system. Or put differently, the body is part of a sensorium, and defined by an interface that is at once human and nonhuman, representational and computational, or sign and signal. Whereas JODI seems to create a distance that reveals the body as almost surreal and allows for a distanced political critique, Scourti's body appears overly intimate (exposed and photographed naked at close range), and yet at the same time, her voice is overly distanced from the object. In this way, her work creates a particular humorous experience of the machine-body sensorium. She shows how the body is read, profiled, and interpreted as a data structure and commercial entity in the metainterface. This writing and reading of the body is a dynamic, continual, and endless stream. Like rapture, it no longer seems to matter whether Google, or any other service providers that employ such scripts, manage to profile the user's body successfully or not. The video displays a strange mixture of her body, the way she is read and profiled, and all the cultural prejudices and commercial models she is subjected to. It thereby illuminates how prejudices, gender stereotypes, and commercial biases are projected back to her (and everybody) even through intimate, bodily interaction.

The body has always been enmeshed in the "new" technologies and economic structures of its time (as indicated by both Lefebvre, Mauss, and Hayles), and *Body Scan* therefore belongs to a history of works that address the (female) body's representation in media. Also broadcast media—like film, TV, magazines, and radio—represent gender in particular ways and have mediated the capitalist consumer subject as part of a "controlled consumption society" (in Lefebvre's terms). As Scourti explains, Export's performance *Tap and Touch Cinema* from 1968, in which male participants are invited to stick their hands into "the cinema" and touch the artist's breast (a box with a curtain attached to her chest), reenacts how the female body is objectified and gendered by media. Though Export subjects herself to the female image, she makes it literal, too. This act of designation is in every way humorous, or as Gilles Deleuze describes humor, it works by way of a designation that destroys meaning: she replaces the idea with the example, behind which there is no hidden platonic "essence."[76]

Whereas Export exemplifies the Hollywood body of advertisement and the movie industry, Scourti exemplifies the body of the metainterface industry. It is a body that not only comes into existence by subjecting itself to a consumer logic that objectifies it (through the movie industry, for instance) but also comes into existence by allowing itself to be read as a signal and be part of a metainterface industry's textual machinery. If the metainterface industry is characterized by a writing machine that transform the HCI of word processing into a global machinery of text—that is, by a reading machine that reads how the reader reads this global text, and by a general control of language that has little interest in what is written, yet great capital interest in the system that produces text and language—then this also transforms the bodily existence of its users: to come into existence, the user does not need to become a particular consumerist body but simply be read as a body. The body in Scourti's *Body Scan* is thus significantly different from the mass media body that is stereotypical and inflicted with sameness. Though arguably mass media still play a role in defining the body as a consuming subject, the neoliberal metainterface body is less dependent on replying to the image of a consuming body; instead, its subjectivity is greatly dependent on the body's readability, which can be measured, calculated, and assessed in a million different ways. As argued by the French philosopher Bernard Stiegler, the individual is "grammatized" and "transindividuated" through what he terms "tertiary retentions" or "technical supports of various kinds."[77] Another concept for these technical supports and their tertiary retentions would be the culture industry.

The New Culture Industry

If the metainterface industry represents a new culture industry, it both resembles and differs from its traditional conceptions. Or put differently, though Lefebvre's critique of a controlled consumption society applies to a contemporary condition, there are substantial differences and additions to be made. As argued above, the metainterface industry not only controls production but also distribution and consumption, and it does so in ways that are not authoritarian. The metainterface's reconfiguration of everyday life does not have a consumerist imperative but rather—through an instrumentalization of writing, reading, language, and bodies—capitalizes on the very being and behavior. The remaining part of this chapter will discuss these similarities and differences in more depth, and furthermore reflect on the current conditions of critique in a contemporary interface culture that bureaucratizes not only culture but also its critique (where critical peer review of cultural software is replaced by protocols and automatization).

In 1944, the leading members of the German Frankfurt school, Max Horkheimer and Theodor W. Adorno, first introduced the term "culture industry." "The Culture Industry: Enlightenment as Mass Deception" is a prominent and often-referenced chapter in their seminal book *Dialectic of Enlightenment*, written in Los Angeles, while they were in exile from wartime Nazi Germany. Given both the Hollywood industry and Nazi Germany's wide use of media to propagate values and behaviors, this seems no coincidence. The chapter describes how a rising culture industry's increasing management of culture is deceiving the masses through a media system of film, radio, and magazines, which is "infecting everything with sameness."[78] In general, the essay is a cutting criticism of the US and increasingly global popular culture that the two German Jewish intellectuals were surrounded by in the booming West Coast metropolis, including film, technical broadcast media, and popular music.

Horkheimer and Adorno describe the growing importance of culture for understanding capitalism; they see culture as an industry that lays the ideological foundation for mass consumption.[79] And their critique is relentless. The culture industry is built around a totalitarian necessity

> inherent in the system, of never releasing its grip on the consumer, of not for a moment allowing him or her to suspect that resistance is possible. This principle requires that while all needs should be presented to individuals as capable of fulfilment by the culture industry, they should be so set up in advance that individuals experience themselves through their needs only as eternal consumers, as the culture industry's object.[80]

In Horkheimer and Adorno's view, the technological development of mass media is propagating a quantifiable sameness through pacifying broadcasts that eradicates individual, reflective perceptions and interpretations. The result is a hegemony of capitalism and consumerism, and an industrialized lifestyle that integrates work and leisure. Fleeing from totalitarian Nazism, they were faced with a totalitarian capitalistic culture industry that manufactured business and consumers as its ideology.[81]

When the computer became a mainstream cultural platform (beginning in the 1970s and 1980s with home computers and gaming consoles), the industrial era was allegedly over. As described by Stiegler,

> A fable has dominated the last decades, and to a large extent deluded political and philosophical thought. Told after 1968, it wanted to make us believe that we have entered the age of "free time," "permissiveness" and the "flexibility" of social structures, in short, the society of leisure and individualism. Theorised under the name of the post-industrial society, this tale notably influenced and weakened "postmodern" philosophy.[82]

In other words, the post-Fordist and immaterial information society and knowledge economy has become part of the theorization of societal developments in the West. Also within cultural criticism, the focus has been more on media theory in the line of Marshall McLuhan's "understanding media" than on understanding the material conditions of culture as an industry.[83]

To many, the modern PC is regarded as a technology to end the standardization of the industrial conveyor belt of capitalism. With the networked computer, industrial standardization is replaced with the flexible customization and modularization of production processes, and leaky borders between producers and consumers broader supply and demand—also coined in the business term "prosumption," and promoted as a "long tail" theory by former *Wired* editor Chris Anderson.[84] Industrial passive consumption is now regarded as more active, "interactive," "participatory," "produsing," or "coproducing." Yet as argued in this chapter, this apparent liberation from industrial standardization leads to controlled consumption through a new metainterface industry.

The metainterface immediately looks flexible, open, and smart to the users. Never has so much cultural content been so readily available on so many platforms, but the amazing efficiency of the new metainterface industry comes at the expense of monitoring, control, and strict licensing. If the conveyor belt produces standard goods in big numbers and is relatively inflexible toward individual consumer demands, the new industry thrives on individual choice, consumption, and coproduction that feed back through a metainterface monitoring every consumer behavior in detail. Agent-based monitoring loops increasingly replace the conveyer belts, and online apps, music, or bookstores and streaming services replace suburban malls. Building a business infrastructure around the metainterface allows for a flexible, fine-grained, and intimate culture industry that can change its public appearance according to demands and trends, while at the same time experiment with new commercial models behind the screen. In this way, the metainterface industry holds the technological infrastructures of an experience economy; it holds the technology that can exploit and capitalize our very experience. With the smart city's urban interface and cloud computing, still new infrastructures are rolled out to monitor, service, and capitalize on behavior (as will be argued in the following chapters). Consequently, Horkheimer and Adorno have become relevant again. As the British digital media theorist David Berry writes, "The challenge for a critical theory of the digital is to critique what Adorno calls identity thinking and a form of thinking that is highly prevalent in computational rationalities and practices."[85] There are, however, also reasons to critique Horkheimer and Adorno's industry determinism.

Metainterface Industry Literacy

To Horkheimer and Adorno, the culture industry system is "never releasing its grip" on the individual but instead renders them "eternal consumers, as the culture industry's object." Resistance is either impossible or will get appropriated. Even "the aesthetic manifestations of political opposites" offer no redemption, but "proclaim the same inflexible rhythm."[86] In popular, industrialized culture, there seems to be no possibility for art to explore a critical tendency. Apart from their criticism, Horkheimer and Adorno's arguments are tainted by an elitist and culturally conservative contempt for the US culture that surrounds them, and often their writing seems like an antidote to Benjamin's dialectical understanding of the relations between art and technological media.[87] Even though it is tempting to disregard Horkheimer and Adorno's points and move on to more postmodern media theory, it would miss the opportunity to understand the current interface industrialization of culture and how it relates to developments in capitalism. There is still reason for critical discussions of the culture industry system, but there is also reason to critically look at Horkheimer and Adorno's understanding of media and technology.

Moreover, Stiegler acknowledges Horkheimer and Adorno as "at once lucid (if not prophetic) and erroneous (if not reactionary)," but he critically points to how Horkheimer and Adorno exclude the retentive dimensions of reproductive technologies (tertiary retention) in their own thinking. They do not reflect that "the very possibility of 'culture,' and thus of 'spirit,' relies on technics."[88] Consequently, their own criticism is also technical—for instance, it presupposes the creation of a critical distance frequently related to writing and academia—but their critique does not reflect on the technological system of writing critique or possible new forms of criticism evolving with (and through) technological activities. As Stiegler points out, in relation to their writing on the phonograph's repetitive reproduction, "Such repetition is possible only through technical recording, only through this technologico-industrial reproducibility that is the objective and infrastructural foundation of Horkheimer and Adorno's analysis of culture industries, echoing Walter Benjamin but, despite Benjamin, *still* failing to think them through."[89] Horkheimer and Adorno fail to understand that new media are not only objects of perception but also integrated into, and thus part of, perception and memory. Cinema develops a cinematic perception, too, or a "kino-eye" (to paraphrase the Soviet filmmaker Dziga Vertov), and Stiegler wonders why Horkheimer and Adorno does not take Husserl and Benjamin into account "ten years after Benjamin writes his famous text, 'The Work of Art in the Age of Mechanical Reproducibility,' whose immense importance completely escapes them."[90] Stiegler argues with Benjamin that reproduction does not just create copies of the real; it adds something constitutive to it: "It is this constitutivity of the technics of reproduction, developed through the culture industries that, according to Benjamin, confer on cinema, for example, its analytic force, which goes beyond its power of alienation—a force that seems to completely escape Horkheimer and Adorno."[91]

Adorno's lack of understanding of technological perception is already expressed in a letter to Benjamin where Adorno comments on the first version of "The Work of Art" essay, and criticizes its positive discussion of popular cinema such as Charlie Chaplin and Mickey Mouse ("dependent art," in Adorno's words), instead of high art such as Franz Kafka and Arnold Schoenberg.[92] In the final, published, third version of the essay, this discussion is cut out, according to Adorno's advice. Yet throughout the essay, Benjamin maintains an interest in "redeeming the reified images of mass culture and modernity for a theory and politics of experience." Hence, the discussion of popular cinema cannot be left out without altering the signification of the essay. It allows, for instance, for a distinction between cinemas, where "certain films might function as a kind of psychic vaccination," and "the cinema becomes an object—as well as a medium—of 'redemptive criticism,'" as Miriam Hansen, a US-based scholar of Benjamin, argues.[93]

Instead of just denouncing technology and media, it can be explored via certain works of art that have a knowledge of their own production, or tendency, in Benjamin's terms. Art is subject to the metainterface industry, but art can also critically explore the metainterface industry. Following Benjamin, the political critique is, however, not just a question of emitting attitudes in the artwork. If art is a probe into the fissures created by technical revolutions, it is not because artists have better insights into society and politics. Rather, it is because artistic production can be a material exploration of its own technological means of production, and how these constantly change as well as create crises in corporate relations, the use of language and copyrights, bodily perceptions, and so on (as demonstrated in the artworks brought forward in this chapter). To Benjamin, media revolutions lead to fissures in art, and by reflecting its medium and technology critically, artistic practice can expose these fissures as a deeper political tendency within the artwork. Critical and artistic practice is hence more related to a material exploration of production and technology, than to an abstract ideology or immediate attitude of the work. Consequently, interface art can be used to delve into a critical understanding of technology informed by material exploration and analysis, which is more nuanced than Adorno and Horkheimer's distanced critical perspective.[94]

Despite regarding the political infrastructures of the culture industry as a totalitarian power that mainstreams and standardizes everything through industrialization, Horkheimer and Adorno also acknowledge that something more fine-tuned and flexible is at stake in the process. At the end of their text, they discuss how the culture industry influences language, and argue that by circulating linguistic models, signification is completed by turning signs into signals.[95] They do not discuss the technicalities, but mainly how advertising, branding, and trademarks influence public discourse, and how ordinary people take up this impoverished language.[96] In a computer semiotic perspective, though, the signalization of signification is carried out by signal–computer interaction in the metainterface (as argued in chapter 1) whose infrastructures are delivered by a rising metainterface industry. The presence of a linguistic system also implies that there still is a grammar at place (following Agre's "grammars of action").

The grammar of the metainterface works through the automated capturing of the everyday behaviors of writing, reading, language, and bodies, but the grammatization does not work automatically. Rather, it is performed through social, cultural, and semiotic processes, which it furthermore performs back to us and reinforces through instrumentalization. Even the code execution in interfaces can be seen in relation to political processes (as the discussion of Nake in chapter 1 pointed out as well). With a reference to Michel Foucault, Chun contends that code execution can be compared to neoliberalism in the way that it stabilizes and privatizes enforcement as self-enforcing law: "from police and judicial functions to software functions," code execution expands judicial interventions and reduces law "to the rules for a game in which each remains master regarding himself and his part." Chun concludes that "this wish for a simpler map of power—indeed power as mappable—drives not only code as automatically executable, but also ... interfaces more generally. This wish is central to computers as machines that enable users/programmers to navigate neoliberal complexity."[97] In this way, Agre, Nake, and Chun all maintain that with interfaces and computation, it is not primarily our immediate language that is influenced but instead our abilities to interact with reality through interfaces (e.g., through their combination of HCI and signal–computer interaction in the metainterface).

To sum up, the assumption in this chapter is that there is a functional basis of the metainterface's production system, and that this industry echoes some of the central features described by Horkheimer and Adorno, including effects on capitalism, culture, and language. Yet there is also a need to consider what it means when the culture industry becomes technologized and digitized through metainterfaces. The postindustrial has been followed by the metainterface-industrial culture as a general paradigm, and this is explored artistically and critically across platforms. The new metainterface industry, at first glance, seems more open and flexible than the old culture industry described by Horkheimer and Adorno, but the efficiency of the metainterface industry comes at the expense of other kinds of control, which are difficult—although not impossible—to address from within the platforms. In opposition to Horkheimer and Adorno, theoreticians such as Benjamin and Stiegler argue for a critical, redemptive potential in certain works of technological art that the artworks presented in this chapter seem to follow. Through the exploration of tendencies, the artworks discussed in this chapter demonstrate how the metainterface industry produces control and language, and how managing systems of language is an essential part of its business model. To rephrase Ong, if letterpress printing and the old culture industry embedded the word and culture in the industrial manufacturing process, making it into a kind of commodity, then the metainterface industry to some extent frees the word and culture again, but by capturing the system it commodifies and grammatizes reading, language, and the user's body in return, as shown in the works by Ubermorgen, Cayley and Howe, JODI, and Scourti. These artworks teach us to read these effects; they train our metainterface industry literacy.

3 The Urban Metainterface: Territorial Interfaces

Spoken Streets

In the app *Las calles habladas*, or *Spoken Streets*, by Clara Boj and Diego Diaz (figure 3.1), the user is offered a random map and walking path around their location. The app seemingly functions like an audio guide, where the user listens to news or facts about places along the way. Unlike most audio guides, however, the narrative appears fragmented. The audio track does not contain the voices of the people living there but rather an automated text-to-speech function. The route planning includes not only a suggested path but also Google searches related to the location. Each user movement generates a search and is answered by the reading of debris from the World Wide Web's enormous body of text: phrases from websites, Twitter feeds, or Facebook events appear together with symbols, numbers, and URLs. Sometimes there is a direct and visible linkage to the user's location, as when the user is near a shop that also is listed on a website. At other times the relation is more abstract, and invites the user to speculate on potential narratives, as when a place is linked to a Facebook event or Twitter feed. On rare occasions, there is only a semantic linkage, as when the name of a street or shop is common, and perhaps located in hundreds of cities. The relation between sense and nonsense, between potential narrative and raving incoherent jabber, seems to be central to the experience of using the app. The confrontation with *Spoken Streets* (i.e., how the web speaks the streets) appears absurd to a human being. It functions almost like a joke or a parody of the ways that sense making functions in other, more conventional applications that, for instance, augment reality by linking locations to the real-time generation of information.

Spoken Streets points to how the city—apart from being a functional entity with housing, factories, offices, streets, electricity networks, sewage, and so forth—can be defined as a "semiotization" of space. From the tourist industry's promotional narratives, personal narratives, and media discourses, to traffic warnings, banners, billboards, street

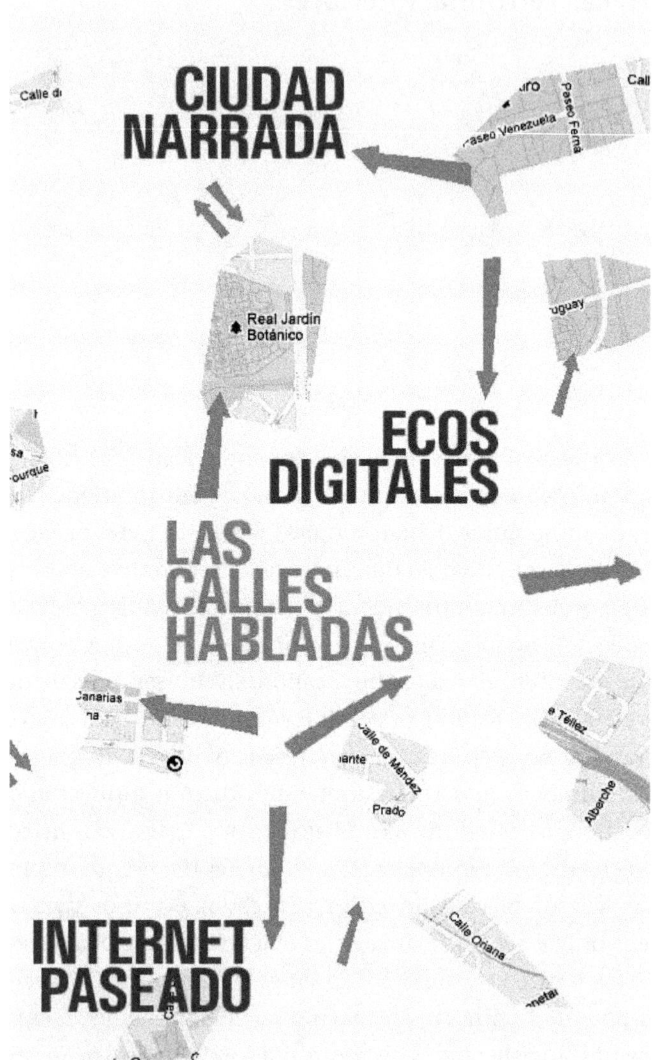

Figure 3.1

Las calles habladas/Spoken Streets (2015) by Clara Boj and Diego Diaz. The app offers a map, a walking route, and an audio guide generated from Internet searchers related to the user's location. Courtesy of Clara Boj and Diego Diaz. Screen capture from app.

names, graffiti, and landmarks, language has always been an indicator of spatiality and power relations embedded in the cityscape. In other words, language is an intrinsic part of an ongoing de- and reterritorialization process within the city. As argued in the previous chapter, the changed production system of writing and reading, and more generally language, represents endless opportunities for publishing, mixing, and sharing tools and services, but it is also a new regime of control that affects the common language activities of everyday life. In continuation of this, this chapter explores how the metainterface is territorial and reaches into space. It looks at how the semiotization of space is affected by a new metainterface industry, and how this disrupts everyday practices of urban life: how the city is read, perceived, inhabited, and organized in new ways.

Examples of everyday urban experiences with metainterfaces are numerous: TripAdvisor provides easy access for tourists to areas, restaurants, and other sights that are otherwise not clearly visible in the urban landscape; with Airbnb, any apartment in the city holds the invisible potential of a bed and breakfast; with Uber, any car can become a taxi and any person its driver; hypes like Pokémon Go give urban monuments or locations new signification (as PokeStops); and so forth. Theoretically, anything can potentially signify anything. All one needs to decode and access the city is the right interface. Or as urban media theorist Martijn de Waal points out, the city is an interface, and seeing this makes it possible to see "how the new interfaces of digital media interfere with this." Particularly, he notices that "every street corner and every local pub leads a double life. The experience of all these places consist not only of their physical design but also of the dozens, perhaps even hundreds, of stories circulating about each place."[1] Social media, smartphone apps, and more, naturally play a vital role in this distribution of meaning today.

In many ways, the disruption of urban space caused by the new location-based services corresponds well with the ways 1990s' net cultures built cultural production on the sharing of resources and disrupted conventional copyright production. The "pirate" net culture is based on an understanding of cultural creativity as a process of exchange, sharing, and reinterpretation, as asserted by Lawrence Lessig, a US professor of law and the founder of Creative Commons.[2] Here, the signification of an image, sound, or text—and now also a place—depends entirely on its contextualization in a remix or mash-up, for instance. Opening up the city to diverse use as well as the sharing of open and free resources—where cars, apartments, streets, places, and so on, can be "remixed," thereby altering their use and meaning—is almost an inevitable consequence of network culture and thus bound to challenge the conventional meanings of a place, tied, say, to its function or history.

Also in urban theory, there are similar discussions about how to develop a more open city. At first glance, the many services that alter the signification and use of the city correspond well with the ideals of urban activists and theorists like Jane Jacobs. Jacobs is world famous for her belief in cities that are open and adaptable to humans' diverse needs and use of a space, and her criticism of urban renewal projects in the 1950s and 1960s.[3] In many Western cities, Le Corbusier's Ville Radieuse and other planned cities for urban life-forms became ideals for a revitalization of urban space after World War II. Old slum neighborhoods were replaced with new areas that in their formal expressions of planning and architecture, favored homogeneity, determination, and predictability. Cities were divided into new residential neighborhoods, office districts, shopping districts, and so forth, all connected by a solid infrastructure for transportation. In Jacobs's view, the urban planners and architects had no understanding of the quirkiness of cities that appears once they are freed from the closed visions. The control and regulation of all elements in the system is, in other words, nothing but an ideal. Faced with everyday use, cities should in Jacobs's mind be designed instead for openness to the chance encounter of a person or place, diverse use, the personal narrative, and so on. In several ways, this is something urban interfaces like Airbnb, Uber, or TripAdvisor often appear to deliver.

In addition, *Spoken Streets* supports these openings of the city and its resources. It invites the user to experience how signification in the urban context is ambiguous, and how this ambiguity can offer openings for alternative experiences that sometimes subvert particular assertions about a place. As much as the urban interface seems to deliver access to a spatial ambiguity that many associate with an ideal urban experience, though, *Spoken Streets* also points to how the means behind it are sealed off and difficult to access. It is hard to get a clear overview of the relations between urban space and the Facebook events, Twitter feeds, and more that the app brings forward. Or to rephrase it, the mechanisms of semiotization that determine how the city should be read, interpreted, and used become increasingly opaque and layered with semiotic processes that are wound up in computation. TripAdvisor, Google Maps, and other apps and interfaces open up the city, but as metainterfaces they are black boxes of signal–computer interactions as well as concealed linguistic and scripted operations.

This chapter focuses on the black box of the urban interface, the urban metainterface, and how it entails a particular territoriality and perception of space. As always (and explained in chapter 1), a usable interface depends not only on its ability to communicate the inner algorithmic processes but also on its ability to capture the user's behaviors. In the case of the urban metainterface, the more the interface opens up the city and becomes adaptable to diversity, the more it needs to monitor the users and their milieu, and process these data. As suggested in *Spoken Streets*, the urban metainterface depends on a combination of location tracking (e.g., the places that the user visits, the people encountered, or more specifically, their devices), and a textual production around these encounters (e.g., reviews of restaurants, shops, or hotels). In other words, the movement of the metainterface industry into urban space depends on a seamless network of sensors (GPS, Bluetooth, etc.) that are embedded into devices and the environment, and a reading of what the users say/write, see/photograph (and ideally also how it feels, tastes, etc.). The production of spatial ambiguity—which is part of the metainterface's ongoing de- and reterritorialization of urban space—builds on network principles where data can flow openly between devices and applications. This allows for numerous opportunities for experiencing and inhabiting urban space in new and "open" ways, but it also holds a certain grammar or "urban gaze"—a particular way of seeing, inhabiting, and making sense of a space.

To unfold this grammar of the urban metainterface, this chapter accounts for how the city itself functions as a process of semiosis that is structured around a particular, often-implicit, and unconscious way of seeing. More specifically, the chapter asks, What kind of gaze does the metainterface produce? It introduces the notion of a "scripted urban space" that encapsulates how the city is structured like an interface (between signification and use), and how the city as an interface becomes increasingly layered with hidden optical and computational processes, such as cameras, screens, and data capture and processing. Many urban art projects have employed media technologies to open up the hidden scripts of the city to human experience and negotiation, and for instance, used screens and cameras to create new social experiences. What is of particular interest in this chapter, however, is how the scripting of space that takes place in the metainterface simultaneously becomes a new *optical* and *computational* gaze of the city. This production of a technological and mediated gaze has historical predecessors. Media technologies have historically always produced particular urban gazes. For example, the panorama and panopticon have played active roles in the construction of the modern urban gaze of the late nineteenth century, and statistics has laid the ground for the gaze of the urban planners following World War II. To circumvent the discussions of smartness and newness that frequently go hand in hand with media-technological perspectives on the city, the chapter focuses on this history, and brings forward artworks that are not entirely new but instead stress the hidden and unconscious optical as well as computational grammars of the urban metainterface gaze. In this way, the aim is to outline a different perspective on the urban metainterface that relates to the histories and politics of urban development and media, rather than the conventional smart city discourse.

Finally, the chapter looks into the cultural logic of the urban metainterface as a mode of production, and how open processes of signification can be employed in urban design. Building on the cultural logic of open-source software and network culture, the metainterface may be seen as a way to open up the city (to access and reconfigure the "source code" of the city). The final part of the chapter seeks to critically account for this cultural logic and, on a more profound level, relate it to a history of "openness" in software culture. Urban design does not, in this view, strive to fulfill an ideal of a particular urban society but rather concentrate on the principle of openness as a way to obtain better societies. In an urban context, open access, open data, open governance, and so on, secure the presence of potential and better alternatives. A smartphone may, for instance, build on open public data to change the signification and use of a place, offer sustainable solutions in the use of a resource, and so forth. Yet the awareness of an optical and computational unconscious grammar in the urban metainterface also suggests a different direction for urban interface design. Insights into the tendency of the urban metainterface may be used to envision alternative trajectories for urban interface design that are more in line with the tradition of "free" software, or an architectural theory that centers not on change and openness itself but instead on how urban societies may deal with change and the ever-present competing alternatives. Inspired by both Jacobs and Richard Sennett, such design strategies favor structures that are open to internal revision according to habitation and a free social moral—rather than the more neoliberal urban development plans that can be associated with openness as an ideal for urban design.

The City as a Text

If the metainterface of the smartphone or the new urban screens can open up for spatial processes of signification, it is because the city is a discursive site. Put differently, if the control of language is a control of space, it is because there exists a fundamental linkage between language and the spatial demarcation of territories. Representation, and more generally language, is not to be confused with the real thing, but it is territorial and creates realities. When it comes to cities, systems of signification are inseparable from the city as a territory. Cities consist by and large of spatial signs. This is their reality: they are made of plazas with statues that are visible signs of institutional power, billboards that are signs of commercial activity, streets that are someone's home or workplace, and so forth. Urban language—for instance, the act of naming streets as well as building (architecture)—is a contestation of urban space, and functions performatively as "speech acts" that declare and control territories.[4]

Roland Barthes was one of the first to reflect on this relation between cities and semiotics. His essay "Semiology and Urbanism" from 1967 begins by evoking Victor Hugo's *Notre-Dame de Paris*, in which the main character, Claude Frollo, the archdeacon of the cathedral, explains the value of architecture to two visitors as a kind of writing in stone. Frollo points to the church with his one hand, and a printed book on his desk with another, lamenting the often-quoted "ceci tuera cela," or "this will kill that."[5] As Barthes notes, *Notre-Dame de Paris* exemplifies the rivalry between two modes of writing: writing in stone, and writing on paper, as two ways of inscribing humans in space.[6]

Barthes is preoccupied with the symbolism of urban space, and sees a conflicting relationship between the functional necessities of a modern city and its semantic charge, which is given to it by history and the way inscription of a metaphysical worldview has taken place in former times (as evident in the architecture of the gothic cathedral). Not unlike Hugo's novel, Barthes's essay has an undertone of urgency to save the architectural city as a locus of signification, which in many ways is understandable at a time of urban renewal, where architecture and urban planning was preoccupied with functionalism, demolishment, and brutalism. As he writes, there is a conflict between signification and reason, "which wants all the elements of a city to be uniformly recuperated by planning, whereas it is increasingly obvious that a city is a fabric formed not of equal elements whose functions can be inventoried, but of strong elements and of neutral elements."[7] The calculative reasoning of statistics that informed urban planners of the city, as explained above, laid the groundwork for many urban renewal projects that replaced neighborhoods of diversity and heterogeneity with homogeneous concrete buildings with uniform purposes. Nevertheless, Barthes's essay may also be relevant today, confronted with the urban metainterface.

In contrast to the urban renewal planners, Barthes believes that a city is formed entirely by processes that relate the oppositions of "marked elements and non-marked elements," meaning signification and the absence of meaning.[8] Put differently, it functions like language and is discursive: "the city is a discourse, and this discourse is actually a language: the city speaks to its inhabitants, we speak our city, the city where we are, simply by inhabiting it, by traversing it, by looking at it."[9] This city-as-language is highly processual and fluent—one person's signified is another person's signifier—and generates a reading that Barthes compares to the systemic avant-garde of the Oulipo:

> The man who moves about in the city, i.e., the city's user ... is a sort of reader, who, according to his obligations and his movements samples fragments of the utterance in order to actualize them in secret. When we move about in a city, we are all in the situation of the reader of Queneau's *100,000 Million Poems*, where we can find a different poem by changing a single verse; unknown to us, we are something like that avant-garde reader when we are in a city.[10]

The point for Barthes is ultimately to save this poetic, expressive, or "writerly" reading of the urban text.[11]

Since Barthes, the semiotics of cities has been widely researched, and Barthes's own elaborations of urban semiosis is also part of his famous *Mythologies*, where he, for instance, elaborates on how the Eiffel Tower is a powerful symbol with no real function.[12] The challenge for an urban semiology is to bring the language of the city out of its metaphoric stage. A haunt for the signified—the real meaning of a place—can, in other words, only be provisional. There is no single lexicon that can account for regular correspondences between signifier and signified in the city, or no "term-to-term symbolism"; rather, the symbols of the city refer to an organization of meaning at a structural level. "Emptying this expression of its metaphorical meaning in order to give it real meaning" is in this sense comparable to the ways in which Sigmund Freud dealt with the language of dreams, as Barthes also notes.[13] With Tokyo as an example, he explains how the signified is a witness to a distribution of signification. From a semantic point of view—and in particular for the Western eye—it is an extremely complex city that lacks a city center as a solid focal point. Yet it is this lack of nucleus that is characteristic, as the empty center organizes the city and gives to "the entire urban movement the support of its central emptiness, forcing the traffic to make a perpetual detour."[14] Or to put it another way, the absence of a fixed signification does not rule out an organizing principle but instead points to the production of signification induced by the reading of the city. The city flourishes with symbolism, where buildings and places have indefinite significances; the city with its lack of stability, its social and discursive practices, is in Barthes's view fluid like a poem—and even erotic in its writerly production of meaning.[15]

If the city holds an unconscious, like Tokyo, as Barthes claims, what is then the unconscious of the urban metainterface that produces signification, one may ask?

One-Way Street

In answering this question, perhaps some inspiration can be drawn from Walter Benjamin, who in his book *One-Way Street*, like Barthes, explores a writerly urban textuality and its relation to other forms of textualities.[16] The book contains a number of rather different short texts organized with street signs, advertising, and other urban texts as headlines. The relations between the headlines and texts are far from clear-cut, and the headlines point to the process of urban reading as opposed to ordering the text of the book. Even though many of the headlines are short, concrete directions and exclamations, they are followed by "thought figures" (or *Denkbilder* in German). In this way, *One-Way Street* demonstrates the writerly expressivity of the urban, but also how this challenges the print tradition. Several of the thought figures are reflections on the relations between text in books and urban text, and how the latter generates new form of reading.

In one of the most cited chapters of the book, "Attested Auditor of Books," Benjamin looks at how his contemporary era is the antithesis of the Renaissance and the situation where printing was discovered: "Script—having found, in the book, a refuge in which it can lead an autonomous existence—is pitilessly dragged out into the street by advertisements and subjected to the brutal heteronomies of economic chaos. This is the hard schooling of its new form."[17] Instead of the autonomous asylum in the printed book, the printed word must now learn from the "dictatorial perpendicular" of film and advertising:

> And before a contemporary finds his way clear to opening a book, his eyes have been exposed to such a blizzard of changing, colorful, conflicting letters that the chances of his penetrating the archaic stillness of the book are slight. Locust swarms of print, which already eclipse the sun of what city dwellers take for intellect, will grow thicker with each succeeding year. Other demands of business life lead further. The card index marks the conquest of three-dimensional writing, and so presents an astonishing counterpoint to the three-dimensionality of script in its original form as rune or knot notation.[18]

This passage has been seen as a prediction of the writing technology of the computer, but it is currently more relevant to view it as a discussion of urban textuality and even a comment relevant to the textuality of the urban metainterface.[19] With the urban metainterface, writing regains its three-dimensionality, but these commercial "locust swarms of print" also change both reading and criticism. Later in the book, Benjamin examines the status of criticism in an age where it is no longer possible to establish the correct distancing and "things press too urgently on human society. … Today the most real, mercantile gaze into the heart of things is the advertisement," he argues, but this does not lead to a simple affirmation of the advertisement or its commercial message, and Benjamin ends the piece with the following rejection: "What, in the end, makes advertisements so superior to criticism? Not what the moving red neon sign says—but the fiery pool reflecting it in the asphalt."[20] This piece is called "This Space for Rent"—a title that points to the space, textual form, and rhetoric of advertisement versus its message, which is absent. The piece literally encourages the reader to enter the rhetorical space of the absent advertisement. In this sense, *One-Way Street* explores how commercialization changes text and reading, and furthermore, argues for a new kind of criticism that enters the urban space and its scripts.

If Barthes induces the urgency of asking what role changes in the production of language (i.e., the interface) potentially mean to the symbolism of the city, then

Benjamin's text is an encouragement to think of this as a critical process that enters the streets. An urban semiology is indeed needed, as Barthes writes, since "if we have difficulty inserting into a model the urban data supplied us by psychology, sociology, geography, demography, this is precisely because we lack a final technique, that of symbols. Consequently, we need a new scientific energy in order to transform such data, to shift from metaphor to the description of signification."[21] To return to *Spoken Streets*, this is what it practices. It is—as a book an action—an interface that enriches the reading of the urban, without enforcing functionality and metaphor. It does not enforce specific ways of reading, by controlling and limiting the expressivity of the city, but opens this process.

As will be asserted in the following, there are two ways of seeing the urban metainterface as a writerly and critical production of signification. First, as an urban interface: like former media technologies such as the panorama, the urban metainterface is part of a "scripting of space" that relates to how the city is experienced and used as the locus of consumption. There are several examples of both artworks and commercial apps that open up the use of the city (or the city as an interface), such as a new visual and scripted public space. The metainterface, however, as indicated by *Spoken Streets*, also has a particular gaze of its own that can be opened to writerly processes of signification. Therefore, second, there is a need to address the gaze of the metainterface and its organizing principle that easily evades attention, but nevertheless structures perceptions and understandings of the city: its unconscious grammar.

The City as an Interface

Rather than accepting that the production of print letters rules out the physical organization of signification (as Frollo, Hugo's protagonist foresees), Benjamin and Barthes induce the idea of the city that functions like language, and whose text is ergodic, or put differently, an interface. As argued in chapter 1, Raymond Queneau and Oulipo build on a literary tradition that breaks with the conventions of the printed book, in which the operational act of flipping the pages is trivial compared to the cognitive act of reading. In Queneau's *Cent mille milliards de poèmes*, reading is a way of figuring out one's way through the text by reconfiguration. In other words, if the city functions as language, it functions as an interface where significations always relate to a textual organization and the reader's own configuration. It relates to a scripting of space that takes place at a material level of organization, infrastructure, and architecture. As demonstrated by *Spoken Streets*, all places are scripted with what Benjamin describes as "a blizzard of changing, colorful, conflicting letters" that controls the access to and understanding of the place, even though not all of these scripts are readable and accessible.[22]

Norman M. Klein suggested the concept of the scripted space.[23] Etymologically, the term "script" refers to something written. It is mostly understood in a literary sense either as a manuscript, meaning the handwritten original, or set of instructions and dialogue for a performance. In information technology and software, drawing on these connotations, scripting refers to the user's own writing of instructions for the computer (e.g., in a Perl script). The meaning of scripts is established in the performance of the script, and not in the coded, incomprehensible instruction itself. Its activation produces something, but the script is rarely read. Behind the affordance of any script (as it may be activated by a user), there is a conceptual framing. For instance, Klein uses scripted space to describe the experience of shopping malls and other spaces that produce illusions though various effects. He compares the experience of the scripted space to the experience of an interface: "an interface may seem invisible or absent, but the audience tends to fill in the blanks, so to speak. The scripting of absence is thus essential not only in architecture, but also in the novel, in cinema."[24]

In the shopping mall, the viewer has the feeling of being the central character in a script. As such, it is a compound that seals off the complexity of urban life and provides a "safe" environment where citizens can indulge in the pleasures of what Henri Lefebvre described as "controlled consumption"—a coming into being through window-shopping or other urban expressions of consumption.[25] Writing the city, by reading the city, is what defines the pleasure of moving about and interacting with places and people, or in Barthes's terms, connecting the "marked elements and non-marked elements," meaning the signification and absence of meaning.[26] The power, however, rests with whoever is able to control the distribution of meaning, or in Klein's terms, the scripting of absence.

Media have played a central role in the distribution of meaning in the city. Hugo's novel significantly changed the meaning of Notre-Dame, just like Charles Dickens's novels changed London, the paintings of Henri Toulouse-Lautrec changed Montmartre, or the films of Martin Scorsese changed New York. Yet media have not only been a central part in the scripting of space as an alternative reading of the city. They are not only an alternative meaning but also part of the textual organization of the city that takes place on the material level of organization, infrastructure, and architecture. Both in relation to the architectural organization of urban space and planning, monitoring, and managing the urban, the city is a highly mediated space.

It is hard to imagine a shopping experience that is not mediated, and various media technologies have throughout history added to the smoothness of mediation. In other words, an interface culture that includes formats that are mobile and locative, and in addition urban screens, media facades, and other integrations with architecture, builds on a longer tradition of mediating spaces. For centuries, media technologies have been used in the organization and perception of urban space (the scripting of space), and specifically in making the reading and operation of the city smooth and pleasurable.

For instance, in London as well as Berlin, Paris, and many other European cities with boulevards, imposing buildings feature characteristic styles of a "grand manner": a totalitarian city where power and appearance is staged and managed through imposing sculptures, buildings, and urban scenery, "a visual language that can bespeak, in proper measure, regimentation, pomp and delight."[27] The architectural style involves the framing of perspectives in the city (the boulevard, plaza, etc.) that are impressive and impose themselves on the city dweller. With the increased urbanization of the Industrial Revolution, the clear framing and perspective was challenged, and this challenge was dealt with through the construction of mediatic miniature landscapes that were manageable on an individual level for the ever-growing urban masses.

The nineteenth-century panorama is an enlightening example of an urban medium that was introduced just before 1800 in major European cities like London and Paris, where it became a popular visual medium that led to many variants and developments, such as the diorama and moving panorama.[28] In its original form, it is a building-size, 360-degree painting of cities or historical events. It creates an illusion of (a lost) overview, but its spectacular nature also becomes part of the city as a designed, media-saturated experience. According to the US urban historian M. Christine Boyer, the panoramic urban form can be seen as a reaction to the loss of overview of the early modern, industrial city—an attempt to contain and control the modern boundlessness or even artificially compensate for its loss.[29] The panorama is an artificial and virtual view in—and on—the city from within a windowless rotunda. Benjamin describes the panorama as an interest to see the "true city—the city indoors. What stands within the windowless house is the true."[30] In this sense, the panorama points to the staging of the urban view as spectacle.

The panorama also directly became part of urban planning. For example, Georges-Eugène Haussmann's grand restructuring of Paris from the 1850s to 1870s introduced large boulevards and enormous squares in order to create an efficient as well as manageable city that conferred to a spatial order "seen from a bird's-eye perspective" that "requested deciphering and reordering" from the street perspective.[31] This panoramic urbanism continued into the twentieth and twenty-first centuries. In 1914 on Potsdamer Platz in Berlin, Haus Vaterland, as another example, offered a Western saloon, Turkish coffeehouse, and Rhine terrace with artificial thunder.[32] Also today, the Sony

Center contributes to the panoramic spectacle on Potsdamer Platz. Likewise, arcades, malls, and even zoos can be seen in the tradition of media-saturated constructed cityscapes where the spectacle compensates for the boundless scale and growth of the city. In the United States today, too, cities such as Las Vegas that are dominated by boulevards can be seen as a continuation of a mediated, spectacular, panoramic urbanity, now designed to be perceived in a cinematic motion through the front screen of the car.[33]

According to media archaeologist Erkki Huhtamo, electronic billboards and other urban projections "constitute a different mode of spectatorship" of an emerging consumer society. The poster is "the modern fresco, and its place is the street," as he also quotes Le Corbusier.[34] As a paradoxical contrast to the modernist aesthetic experience of immersion (into the book, painting, or darkness of the film theater), posters, billboards, and projections present an aesthetics of distraction more suited for the flaneur. The aesthetics and politics of urban space are clearly visible in the distractions of flânerie, but commodity culture operationalizes and habitualizes it, and turns the constant (and empty) distractions into productive experiences oriented toward consumption, as already pointed out by Benjamin's reflections on advertisement (described in *One-Way Street*). *The Society of the Spectacle*, as Guy Debord labeled it, is a conglomerate of flickering images (TV channels, billboards, etc.) that fetishizes consumption and absorbs people. Debord's counterstrategy is the *dérive* (drift), which can be understood as a continuation of the erotic reading strategies of Barthes and Benjamin (but as discussed later in this chapter, it is a counterstrategy that easily lends itself to instrumentalization).

Subsequently, media technologies not only alter the signification and symbolism of cities but also affect the ways cities are organized and configured by their inhabitants: they alter their interfaces. The society of the spectacle has, for instance, changed urban space as a location for civic public engagement. Public space is out-conquered by not only billboards, posters, and other urban images but also the public image and sound of television and radio, consumed in private at home. For many people, the image of the Western suburbia of the 1960s and 1970s is inherently linked to empty streets and the consumption of broadcast media. Today, media facades, locative media, sensors, and pervasive computing offer ways of configuring the use of a city: they open the city as an interface. As media theorist and urban historian Scott McQuire contends, "With the emergence of mobile devices, media consumption is increasingly occurring in public. It could be argued that the 'media event' is in the process of returning to the public domain."[35]

The following parts of this chapter will argue that McQuire's assumptions are correct, and that much urban media art provides illustrations that lead the way to a more writerly and open city. This openness of the city as an interface takes place on two levels. First of all, it is a visual openness, in which cameras and screens play an active role in the creation of a new urban public sphere. Second, it is a scripted openness, in which instructions and algorithms reconfigure the experience of the city as a shared space. More important, however, both the visual and scripted/algorithmic openness hold an unconscious layer. This means that as much as there will be interfaces that open up the city and make it writerly (and often artistic uses of interfaces serve this purpose), there will be a need to open up the interface itself and its intrinsic production of meaning as well. In other words, within the urban interface itself there are organizing principles that appear absent and unconscious, but nevertheless distribute meaning: the urban interface is not just an interface to the urban but also a metainterface that conceals scripted processes.

The Optical Openness of the City

Screens, as visual dimensions of the urban interface, immediately catch the eye, and many artists have drawn on experience and knowledge from network culture and new media art to reconsider the functioning of its realism. In an account of the first Urban Screens conference held in Amsterdam in 2005, urban theorist and curator Mirjam Struppek writes on the potential of screens for urban communities that "interactive screens integrated into urban furniture can help to circulate and access data for comments, stories, or conversations that characterize and strengthen the local community."[36] Likewise, McQuire argues for the "need to imagine uses of screens which don't simply … produce alternative content, but are directed to producing new forms of public relationships."[37] Both Struppek and McQuire, along with many other theorists of urban media, have suggested creative and artistic strategies from the sphere of electronic networks as a conveyor of new social relations in an urban space saturated with new media technologies.

Struppek refers to historical installations, such as Sherrie Rabinowitz and Kit Galloway's *Hole in Space* from 1980, where they connected the front of Lincoln Center for the Performing Arts in New York City and a department store located in the open-air shopping center of Century City in Los Angeles with two life-size, live television projections. But not least she provides numerous examples from the early 2000s, where the Internet becomes "a delivery mechanism to inhabit and/or change actual urban spaces." For instance, Karen Lancel's *Agora Phobia* (2000–2004) deals with the personal consequences of urbanization along with issues of isolation and loneliness. In an attempt to open up social interaction, Lancel invites participants to chat with "people who were isolated and feeling like prisoners, nuns, asylum seekers, or 'digipersonas.'"[38] Likewise, in the work *Karlskrona2* (1999), the artist group Superflex along with CAVI from Aarhus University created a virtual copy of the Swedish town Karlskrona that was accessible via the Internet. Through the replica, the group claimed, "things will change, buildings will redefine their function, social hierarchies will alter, laws will be reconstituted and renewed."[39]

As McQuire remarks, the relational architecture of Rafael Lozano-Hemmer is perhaps one of the best examples of how artistic work may employ large screens to redefine the spectacle of the media-saturated urban space as something other than a relay for live sports or billboard advertising. The work *Body Movies* (figure 3.2), for instance, is an installation first shown on the Schouwburgplein in Rotterdam in 2001, and presented by the V2 Institute. *Body Movies* automatically displays thousands of photo portraits on the facade of the Pathé Cinema building, and thus creates a facade that stands in

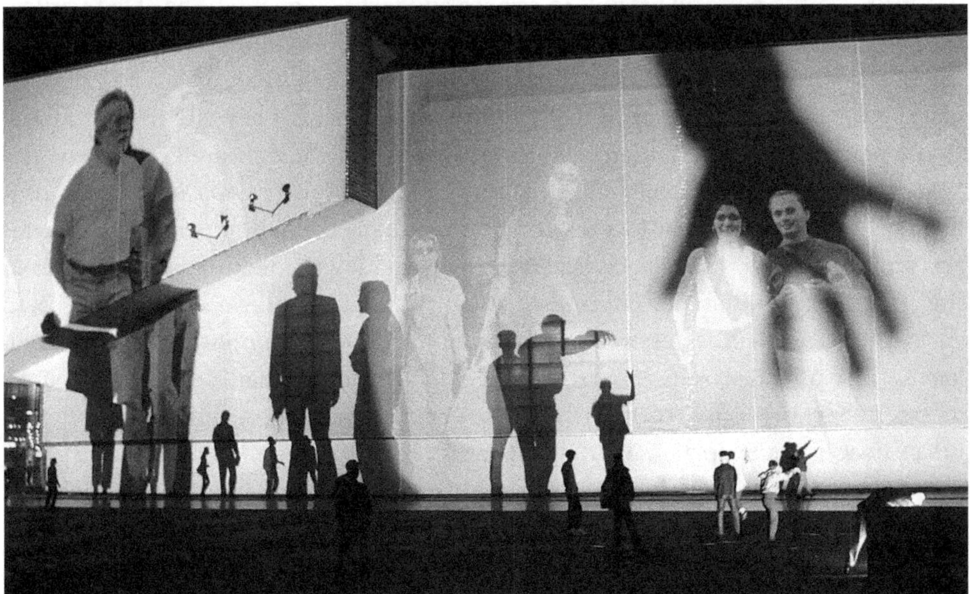

Figure 3.2
Rafael Lozano-Hemmer, *Body Movies, Relational Architecture 6*, 2001. Schouwburgplein, V2 Cultural Capital of Europe, Rotterdam, The Netherlands. Thousands of photographic portraits are displayed on the building (Pathé Schouwburgplein in Rotterdam) using robotically controlled projectors that are activated by by-passers, and appearing in their silhouettes. The installation is an example of how artworks redefine the spectacle of the media saturated urban space as an interface to the city. Photo by: Jan Sprij.

heavy contrast to the spectacular representational commerce of the building's inside. The building is almost transformed into a precinematic shadow dance where the photo portraits appear inside the silhouettes of passersby, created by strong light projections. Lozano-Hemmer's work is a playful interaction with the urban environment, and what Timothy Druckrey calls an "anti-spectacle" that contrasts the immersive cinematic experience of visual effects.[40] Rather than creating passive cinematic mass consumption, Lozano-Hemmer demonstrates how technology can perform as an interface that, in his own words, develops "a unique relationship with a distributed public without losing sight of either identity, locality, or with the delicate meanings of interactivity": an "intimacy within an intimidating scale."[41]

Lozano-Hemmer, and before him artists such as Galloway, Rabinowitz, and others, all demonstrate how cities can be opened up by the "avantpreneur's" alternative use of technologies. Galloway and Rabinowitz introduced the term avantpreneur in the1980s to describe the practice of their project, Electronic Café International. As they explain it, "artists alert to emerging trends in science and technology might be able to articulate the intrinsic qualities and dangers of unclaimed territory not yet targeted for total exploitation by the entrepreneurs."[42] It is also clear, though, that screens and cameras have become an intrinsic part of today's urban experience for any city dweller as well as the practice of any urban entrepreneur; screens and cameras are everywhere as a "panoramic" spectacular experience. Galloway and Rabinowitz's *Hole in Space* has now literally been transformed into a smartphone app that along with many other cameras and screens (large and small), takes part in the formation of identities and localities in the city, as Lozano-Hemmer imagines. Naturally, this is not only a spectacular experience. The screens and cameras also impose a ubiquitous "panoptic" state of surveillance. Cameras and screens are part of a surveillance scheme that is driven by a public interest in security as well as efficiency. Likewise, the production and consumption of urban images (on Snapchat, Twitter, Facebook, etc.) that monitor people are part of social media's economy, which not only provides a new panoramic experience but is based on the reconfiguration of a panoptic surveillance scheme, too.

The panoramic and panoptic dimensions point to how the screens and cameras hold a particular gaze within them. The artist employment of the interface may serve to open up the city as a writerly text, and explore the pleasures of the chance encounter, reflection of images, or reconfiguration of a building. It may, however, also explore the panoptic and panoramic as a writerly interface itself; in other words, how screens and cameras not only capture and project images of people and places but also distribute meaning. This is a quest for the unconscious principle of the metainterface that gives rise to other technological pleasures and anxieties.

Panorama-Panopticon: *Faceless*

As exemplified by the panorama, the relation between monitoring and aesthetics is far from new. For instance, Jeremy Bentham's famous panopticon is an example of an inverse panorama where the prisoners have swapped places with the images, and the guards with the audience. As an influential scholar on the panorama, Stephan Oettermann precisely phrases it: the panorama and panopticon "are at the same time identical and antithetical: in the panorama, the observer is schooled in a way of seeing that is taught to prisoners in the panopticon."[43]

The panorama is a product of the same society as the panopticon; they are two sides of a media architecture that, as Oettermann explains, is preoccupied with the organization, control, and experience of the masses. The panorama, as an aesthetization of this, does not simply construct the gaze of and at the masses but instead points to the persuasive beauty of the event itself: the dream of a socially sublime perspective where everything can be observed at once. It gestures to the desire of assuming the position of the omnipotent voyeur as well as the exhibitionist who is the object of the all-seeing gaze. Such desires are not simply instrumental but rather part of a compliance between instruments and humans, and are a necessary means to understand the gaze of the metainterface, present in social media such as Instagram as well as Google Street View panoramas, photospheres, and virtual realities.

This desire to be part of the urban social interface—being visible to the network—is increasingly becoming part of a process of individuation (as also observed in the previous chapter, in relation to Erica Scourti's work). People willingly log onto networks that monitor their location, use of data, and so forth. In light of totalitarian regimes, this seems inexplicable—as if all users are blinded and do not see the consequences, as if they are unable to control their narcissistic, exhibitionist, and voyeuristic desires. Yet these behaviors—the exhibitionist's desire to submit to surveillance, and the voyeur's desire to see—should be seen as two sides of the same coin: in each case, they reflect a particular politics and production of space, a particular media-technological grammar.[44] This politics of space, associated with the panoptic/panoramic aspects of interface culture, is made perceivable through a number of artworks and happenings that demonstrate a new grammar of urban space, brought about by the city's new visual information infrastructures.

Numerous artists have addressed the panoptic aspects of modern cityscapes—which since September 11, 2001, have evolved continuously with the proliferation of urban terrorism. One of the most sophisticated cinematic experiments with surveillance cameras is the fifty-minute feature film *Faceless* by the Austrian, British-based artist Manu

Luksch.[45] *Faceless* is shot in London, a city reputed for its high density of surveillance cameras.[46] The film entirely consists of recordings from surveillance cameras in the city. To gain access to the video feeds, Luksch took advantage of the legal framework around rights to individual data. In various ways, this becomes a central theme in the film. If the data are personal (i.e., the person appears on camera), it is a larger operator that applies numerous cameras, and all other actors are anonymized, then according to the British Data Protection Act of 1998, any person has the right to buy copies of data where they appear, for a maximum total cost of ten pounds. Consequently, apart from the one requesting access, all persons appearing on the recording will have their faces blurred, masked out, or in other ways erased.

In short, *Faceless* is the story of a woman (played by Luksch herself, according to the legislation) who lives in a future, technological dystopia. It is a world where "The New Machine" controls time. Time is no longer organized chronologically and continuously but instead as pulsating moments, or so-called RealTime. This machinic operation of time controls the lives of all citizens, and pervades their consciousness to the extent that they block out all memories and expectations of a future, and hence also their feelings of regret, guilt, fear, and anxiety: "RealTime, the perfect and perpetual present, is the heartbeat of the healthy universe," as the narrator summarizes it.[47] Moreover, The New Machine functions as a perfect Big Brother system that monitors the citizens' traces of data, and sends out "Overseers" to correct errors and discrepancies.

The female protagonist, employed as one of the thousands of people who analyze data, is "haunted by an echo of a memory."[48] She receives a letter that encourages her to follow her dreams, and pursues them across the city. In one scene, she gets her own face back, but is then chased by Overseers and encounters a group of "spectral children," marked by colorful spots on their faces, who have been liberated from The New Machine. Later, she meets her ex-lover, who explains the real cause of events. He was the one who sent her the letter. He has worked as an Overseer, but offers to smuggle her with a "Seapod" across "The Blue Water." She then succeeds in infiltrating The New Machine and liberating the city, and is reunited with past and present, her lover and their child. The narrator, however, finishes the film by inducing a moment of uncertainty: "Her *child*? Her *lover*? Doubts multiply. Is this *her* dream, *her* past? Or nostalgia for a time that never was? A prison of another perfect present?"[49]

Dehumanized Ready-mades: *Faceless*

The narrative follows a classical Big Brother scheme that repeats a criticism of surveillance in the tradition of George Orwell's *Nineteen Eighty-Four*.[50] First and foremost, the film is a criticism of a contemporary society where monitoring and control has become an accepted everyday practice that can deal with anxieties of the future as well guilt of the past, and where the loss of freedom happens with general consent.[51] In its exploitation of the mechanisms of surveillance, the film expresses a strong criticism of not just the legal framework around visual surveillance but also how it is handled in an everyday practice. Luksch has processed these legal conditions into a "Manifesto for CCTV Filmmakers." The manifesto is a creative guideline where the filmmaker is asked to "consider the visual impact of this manipulation, and to establish a rule for the handling of footage delivered with ineffectual masking or blurring."[52] The main rule is that the recordings must be authentic, and there is no use of additional lighting or camera settings. *Faceless*, in other words, is made entirely from raw surveillance recordings obtained through the Data Protection Act and used to explore the law (figure 3.3). Luksch and the film's composer, Mukul Patel, explain how the film has evolved over the years: "There was no traditional shooting script: the plot evolved during the four-year long process of obtaining images. Scenes were planned in particular locations, but the CCTV [closed-circuit television] recordings were not always obtainable, so the story had to be continually rewritten. *Faceless* treats the CCTV image as an example of a legal readymade (*objet trouvé*)."[53]

The planning of *Faceless* evolves as a constant response to the circumstances and hence demonstrates how (despite the Data Protection Act) many surveillance operators are reluctant to allow access to their recordings. Obviously there are significant costs involved in the anonymization of faces, but Luksch also accounts for a number of instances where the cameras did not work, and in fact had never worked.[54]

Nevertheless, the plea to consider the visual impact of surveillance cameras turns the film into something more than the usual Big Brother criticism. It holds a certain melancholic beauty that is not only reinforced by the soundtrack (by Patel) but also by the subtle attention to the materiality of video surveillance. This material beauty may be seen, for instance, in the recording glitches that visualize the breakdown of The New Machine's panopticon in the narrative. The aesthetics of surveillance cameras is emphasized by a lack of focus, dirt on the camera lens, overexposed images, saturated colors, and so on. The positions of the cameras, too, with their fixed immobility and more or less random bird's-eye perspectives, are characteristic. More generally, it is a film that aesthetically reflects the politics of time and space in a city controlled by the information infrastructure of surveillance cameras. The CCTV recordings have a certain dehumanized and objective quality, which also influences the organization of time and perspective in the film.

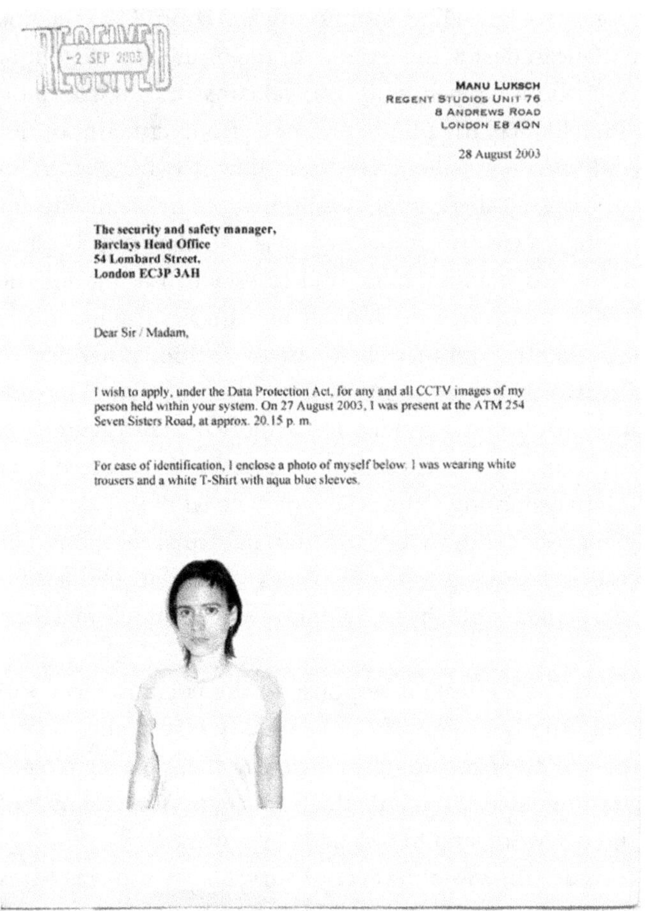

RECEIVED
-2 SEP 2003

MANU LUKSCH
REGENT STUDIOS UNIT 76
8 ANDREWS ROAD
LONDON E8 4QN

28 August 2003

The security and safety manager,
Barclays Head Office
54 Lombard Street,
London EC3P 3AH

Dear Sir / Madam,

I wish to apply, under the Data Protection Act, for any and all CCTV images of my person held within your system. On 27 August 2003, I was present at the ATM 254 Seven Sisters Road, at approx. 20.15 p. m.

For ease of identification, I enclose a photo of myself below. I was wearing white trousers and a white T-Shirt with aqua blue sleeves.

Figure 3.3
Faceless (2007) by Manu Luksch. The movie is made of recordings from surveillance cameras. Access to the footage is permitted through the legal framework around rights to individual data. Courtesy of Manu Luksch. CCTV request letter.

First, time is a central theme, not only in the narrative, but also in the material organization of the film. Time is organized into one image per second (instead of the usual twenty-five images per second), and this is reinforced by the beats of the soundtrack as well as the ever-present time stamps on the images. Luksch explains how RealTime was developed in response to the ways the audience focused on the time stamps during test screenings of the film.[55] As with the main topic of facelessness, the use of time in the narrative is a direct response to the character of the recordings. Time is counted and speeded up, thereby demonstrating processes that are usually not perceivable. For instance, the passage of shadows over a landscape or tidal waters are condensed into passages of a few minutes. At other moments, time is paused or looped. Time, in other words, becomes something that can be manipulated and controlled through the machine; it is completely detached from duration and the lived experience of the citizens, and yet it controls their lives. The machine controls time, but that does not lead to any insights or provide any narrative coherence for the individual.

Second, perspective is equally remarkable in the film. It is a camera perspective that is far from the subjective camera conventionally used in films. Instead of portraying the individual (e.g., the protagonist), the bird's-eye perspective of the surveillance camera is aimed at capturing as much of reality as possible in one image (figure 3.4). The surveillance cameras produce an automated, all-encompassing, and objective visual

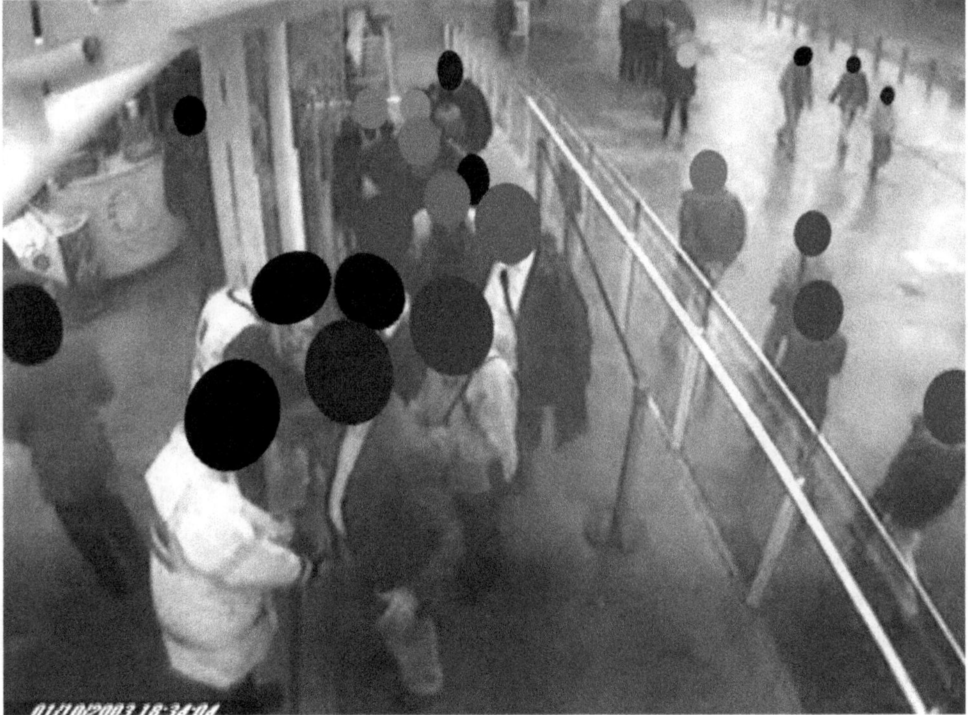

Figure 3.4
Faceless (2007) by Manu Luksch. According to the British Data Protection Act, access to surveillance footage is dependent on the erasure of all faces other than the person requesting access. This becomes a central part of both narrative and visuals in Luksch's movie. Still from movie.

perspective of the masses. The film is a representation of a surveyed space that presents how the cameras capture the whereabouts of any citizen in London, and how it could look—if only they had access to the cameras. In this sense, the film is an expression of realism, where the aim is to present how the world is produced (rather than what it is), and not least, the politics of this production reflected as a question of access to the movement of the masses.[56]

Mass Ornamentation: *Faceless*

According to German cultural critic Siegfried Kracauer, the film seizes the movement of the masses as a "mass ornament" (figure 3.5). In 1927, Kracauer analyzed how the modern dance show and stadium spectacle becomes a meaningful modern experience.

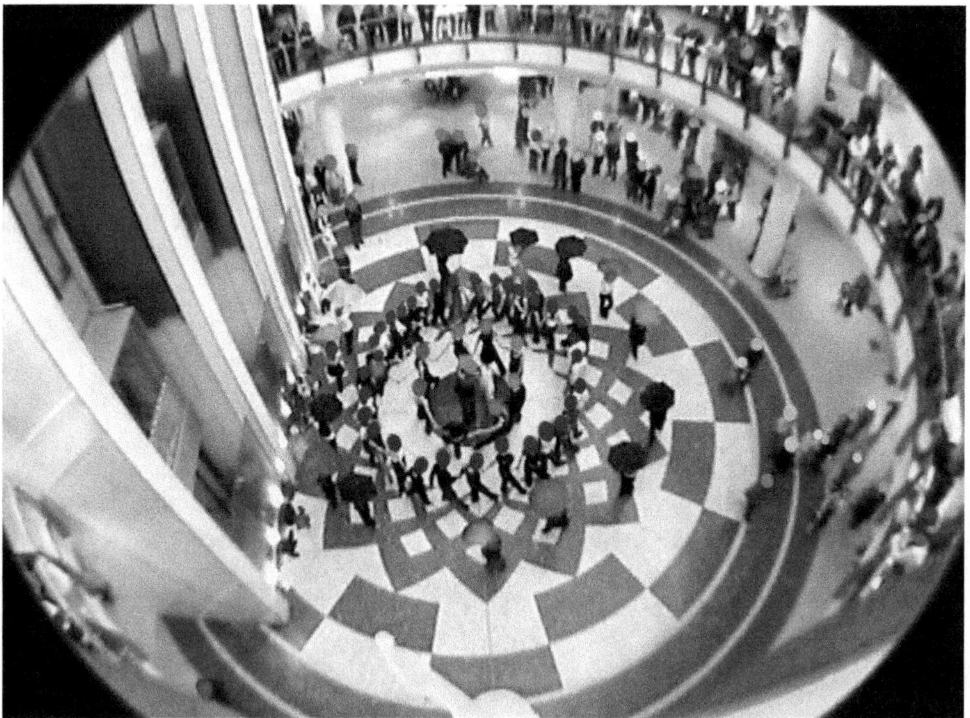

Figure 3.5
Faceless (2007) by Manu Luksch. The movie includes the *The Eye—Choreography for surveilled space* (2005) where 80 dancers perform in front of a shopping mall's surveillance cameras. Luksch has referred to this as a tribute to the choreographer Busby Berkeley's panoptic camera shots of dance parades in the 1930's—an example of what Siegfried Kracauer has also called a "mass ornament." Still from movie.

In his understanding, the spectacle (as a mass ornament) is an aesthetic reflection of capitalism's production process, the assembly line, and statistical control. All these characteristics of a modern mass society are generally abstract or invisible, and can only be recognized by the individual as an indirect experience, or staged aesthetically as a mass ornament: "The production process runs its secret course in public. Everyone does his or her task on the conveyor belt, performing a partial function without grasping the totality. Like the pattern in the stadium, the organization stands above the masses, a monstrous figure whose creator withdraws it from the eyes of its bearers, and barely even observes it himself."[57]

In this way, mass society is experienced and reflected through the mass ornament. But at the same time, the spectacle enchants the mass society: it is experienced as a common project that binds people together. As Kracauer writes, "The bearer of the ornaments is the *mass* and not the people."[58] The mass ornament exists above the level of the individual. This enchantment is what fascism and Nazism took advantage of, or rather, misused and made perverse, as seen, for instance, in Leni Riefenstahl's films of the period.[59]

Though *Faceless* does not explicitly draw on Kracauer's analysis of dance shows, it includes the staging of *The Eye—A Choreography for Surveilled Space*, where eighty dancers perform in front of a shopping mall's surveillance cameras. Luksch has referred to this as a kaleidoscopic and humorous tribute to the choreographer and musical director Busby Berkeley, who in the 1930s became renowned for his panoptic camera shots of dance parades that revealed the patterns of the dance and made the dancers part of a larger superorganism (as ants in an anthill, or humans in the machinery of mass society).[60] In this choreographed part of *Faceless*, it is clear how the monitored dancers become patterns, which is also emphasized by the colored dots over their faces. In other, less choreographed and more random scenes in the film, the dehumanized mass perspective is created by the fixed CCTV camera's lack of intention.

The Optical Unconscious: *Faceless*

The aesthetically pleasing element of the film is hence not just its dystopic atmosphere and critical voice. It is also its attention to a really existing superhuman construction of time and space, which is addressed in the narrative, but more so revealed (as a realism) in the visual aesthetics along with the use of the visual materials of time and space as "legal readymade." In several ways, *Faceless* thus expresses an artistic resistance to complete surveillance by making an "optical unconscious" perceivable, or in the words of Benjamin, it makes it tangible that there is "another nature, which speaks to the camera rather than to the eye. 'Other' above all in the sense that a space informed by human consciousness gives way to a space informed by the unconscious."[61]

Benjamin's notion of an optical unconscious suggests the need for a kind of photographic analysis of the modern, urban world. He introduces the concept in "Little History of Photography," where he also discusses the Parisian pioneer of documentary photography Jean-Eugène-Auguste Atget, who photographed the streets of Paris on the edge of modernization. As Benjamin writes, Atget "looked for what was unremarked, forgotten, cast adrift" and made photos that "suck the aura out of reality like water from a sinking ship." In other words, the optical unconscious of photography is about seeing the modern city in a modern way, set free from romantic embellishment and related to a massified perception. This is a kind of perception "whose sense for the sameness of things has grown to the point where even the singular, the unique, is divested of its uniqueness—by means of its reproduction."[62]

The optical unconscious is marked by the industrial reproduction of sameness, which paradoxically becomes visible through the nonhuman optics of the camera. But whereas many understandings of photography have focused on the visual, indexical realism of photography, Benjamin argues that photography adds new forms of inscription and "literarization" of the urban, too, and that this changes the way of seeing the city through its optics:

> The camera is getting smaller and smaller, ever readier to capture fleeting and secret images whose shock effect paralyzes the associative mechanisms in the beholder. This is where inscription must come into play, which includes the photography of the literarization of the conditions of life, and without which all photographic construction must remain arrested in the approximate. It is no accident that Atget's photographs have been likened to those of a crime scene.[63]

If Kracauer's concept of the mass ornament describes the social dimensions of the mass society, Benjamin's notion of the optical unconscious demonstrates how it influences the way of seeing things, objects, and urban environments as a "crime scene" calling for inscription and literarization. This of course relates to how CCTV also "criminalizes" urban space as a setting ready for something to happen and the city as text (discussed above in relation to Barthes and Benjamin): it outlines how the optical and textual might be related through a nonhuman perspective that calls for inscription.

Both the mass ornament and optical unconscious can be exploited as control or spectacular enchantment, but exploring its ways of working, and even its fascination, may lead to critical appreciation as well. *Faceless* thematizes how urban space has become monitored and captured in databases, but it also uses this material in its cinematic aesthetics. Surveillance is traditionally a hidden, one-way gaze, and the monitored person has basically no opportunities for interaction.[64] Often, CCTV recordings are missing, impossible to find, or do not show what one is looking for. An important point of Luksch's film is all the dysfunctional images and lack of transparency. In this way, *Faceless* highlights that the total panoptic view does not exist, or at least it has severe technical difficulties, and the film's surveillance plot mainly appears displaced in the gap between the actual images and perspectives.

Faceless shows both the story of invisibility, the never established all-seeing eye, and the phenomena that become visible between the images, perspectives, and cameras. The images of *Faceless* are not just pictures of monitored subjects; they are also pictures from (and of) the optical unconscious of urban surveillance—a society based on surveillance, where the dehumanized perspective of surveillance is an integral part of what keeps society coherent. It reveals a surveillance-based mass ornament. Surveillance is no longer something we can choose to turn on or off; it is integrated into the development of society as an urban metainterface culture. Surveillance and the urban metainterfaces are new stages for performance, or to recall Barthes, new elements in the urban language. *Faceless* transforms surveillance into a way to sense the invisible, abstract structures that make up the city and societal machinery—abstract structures that each of us is only a small part of while not being able to recognize the whole, as Kracauer argued. Through the way it explores the production of surveillance, *Faceless* generates a visual aesthetics that instead of hidden monitoring, investigates a critical mass ornament. *Faceless* examines a visual dimension of the urban metainterface, its gaze, including how this gaze is configured by optical unconscious infrastructures and simultaneously allows for datafied control.

The Opening of the Urban Script

The ways cities are used (or "configured" as an interface) are prescribed and scripted. The camera and screen have central roles in this scripting of space. They make, for instance, the city both spectacular and panoramic, and the locus of panoptic surveillance. Urban cameras and screens, however, may also be employed differently and take an active part in a reconfiguration of the urban experience. They may make the city writerly, and open up the meaning and signification of places, people, and events. As seen in the works of Galloway and Rabinowitz, Lozano-Hemmer, and many others, this openness is constructed visually, but the construction itself points to an optical unconscious of the new urban masses (as seen in *Faceless*).

Crucially, though, the scripting of space does not only work visually. Many urban theorists have pointed this out, including Lefebvre, who famously asserted that urban space cannot be understood as a simple accumulation of objects and people but rather as a societal practice that may be perceived, lived, and reflected.[65] As also explored by the Situationist movement in the 1950s and 1960s, the city has an ideological organizing principle, which defines its spaces and spatial practices. As elaborated by Simon Sadler, Situationism can be seen as a direct response to, for instance, the strict zoning laws (for housing, work, and leisure) and traffic circulations that characterize the functionalist modern city. Situationist practices strive to dissolve the boundaries of the urban zones, such as the boundaries between the public and private, or work and leisure, and make the adjoining moments, the travel and transportation itself, pleasurable.[66] Following from a focus on the optical and visual construction of meaning in the city, there is a need for a similar critique of the operating principles of zoning and transportation. Sadler provides an extensive elaboration of Situationist strategies and thinking, although similar attempts to deconstruct urban space and spatial practices have been employed through urban interfaces.

One such example is the work *Serendipitor* (figure 3.6), a smartphone app produced in 2010 by artist, architect, and researcher Mark Shepard.[67] It reads the position of the user on a map to create a journey planner for pedestrians. More specifically, it uses the data to subvert the traditional smartphone route planner by instructing the user to, for example, "find something on the square nearby and photograph it," or "look for someone who is lonely and ask to walk with them for a while" (as it is proclaimed in the app).

If conventional mobile journey planning is about getting the user aptly to a location, *Serendiptor* draws on a history of mapping developed by the Situationists. The psychogeographer explores "the precise laws and specific effects of the geographical environment, whether consciously organized or not, on the emotions and behavior of individuals," by letting the drifters be guided by "the attractions of the terrain and the encounters they find there," as explained by Debord.[68] Situationist strategies in this way open up the compounds of the city that seamlessly guide the citizen and their way of seeing; it opens up to the absence that is otherwise hidden in architecture (an unconscious scripting of absence, as Klein put it above) or the excessive reading argued by Barthes.

Whereas psychogeographic drifting is exposed to the randomness of the environment and one's individual mood, Serendipitor also exposes the user to the environment of the app—that is, its predefined instructions and functionalities. In other words, as a Situationist journey planner, it points to the programmability of "the dérive." The letting go and playfulness of the dérive is not only countered by the restraints of the topology of the urban environment but also the topoi of the program itself. It instructs the user in various ways ("find the nearest tree and sit under it for one minute," etc.),

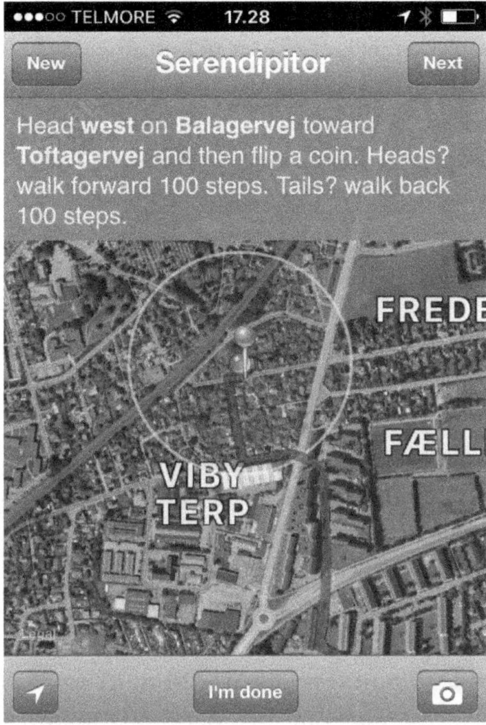

Figure 3.6
Serendipitor (2010) by Mark Shepard. A smart phone app and interface to the city that applies situationist strategies of mapping to Google's routing service. Courtesy of Mark Shepard. Screen captures from app.

and more fundamentally demands of the user to be instructed. In many ways, one could claim that Situationism does not deny the spectacle of mobile technologies but instead almost lends itself too well to the instrumentalization in an app interface. The appearance of Endomondo's #TrackYourArt Contest and similar phenomena further underlines this.

In other words, replacing the effectiveness of a journey planner with the dérive or the spectacle of the urban screen with relational architectures are both invitations to explore the significations of the city, open the city up as an interface, and experience an alternative scripting of space than the one induced by everyday use as well as functional metaphors. Urbanization is in many ways a both seductive and alienating process, but the creative and artistic use of cultural logics developed by network cultures can help open the city up in new ways. Nevertheless, it also points to the interface itself as an urban metainterface. The interface holds an alternative topology that is made up of different connections and compounds than the Euclidean space of both the space of the screen and journey planner. As much as Situationism may open up the scripts and configuration of the city *as* an interface, these configurations depend on a logic that (following Klein as well as Benjamin) appears absent and unconscious—as a hidden grammar or tendency within the interface *to* the city itself.

The Computational Unconscious: *London.pl*

Perhaps it could be argued that with the metainterface industry, the mass ornament of the stadium dance show (which Kracauer described) has been replaced with hordes of Pokémon Go players who move about the city subject to the more or less random appearances of creatures in a game. In other words, as much as there exists an optical unconscious that speaks to the camera rather than the eye, there is a reality of numbers and data that speak to the machine as opposed to the human perception: a data unconscious. The surveillance cameras in *Faceless* show a visual perspective of the masses, but this visual perspective is parallel to the more abstract information perspective of data surveillance. In *Faceless*, this informational perspective on the masses is extracted through the data traces collected by The New Machine (the protagonist is occupied as a data analyst). The visual perspectives of urban interfaces are, in other words, followed by an attempt to explore the absent infrastructure, grammar, and datafication. Others have explored this dimension, such as British artist Graham Harwood. Harwood's work *Perl Routines to Manipulate London*, or *London.pl*, shows this other way of seeing London: the unconscious of computation and data, which is at once fascinating, seductive, and contributes to the panoptic prison of bureaucracy and capital—but is also poetically transgressive and liberating.[69]

London.pl is an example of a Perl poem (poetry in the computer language Perl) that with reference to the classical poem by William Blake, relates the sensuous experience of the city to an ontology of programmability and computation. In Blake's poem *London*, the protagonist wanders through the streets of the city, observing and describing the despair of the citizens. In a rush of indignation and fury, the poem expresses a human perception of class oppression that led to the revolution in France, and was expressed politically in Thomas Paine's *Rights of Man*, published not long before. To put it another way, Blake demonstrates how class oppression can be understood not only logically but also aesthetically, by means of a sensual cognition of the city. Not least, the perception of sounds plays a central role: the sigh of soldiers, cursing of prostitutes, and cries of men and infants become direct marks of economic, social, and political oppression that follows from the "chartering" of urban life. The only thing that is not chartered, or made proprietary (the only thing left, so to speak), is the children's cries of fear.[70]

Harwood's *London.pl* from 2002 is perhaps not as renowned as Blake's poem, but Florian Cramer calls it an "outstanding example" of Perl poetry.[71] Perl is a high-level scripting language, and although it has sometimes been criticized for being a "write-only" language that is too complex for other programmers to read because of a convoluted syntax, it is widely used for code poetry. Harwood's extensive annotation also makes the poem quite understandable to programming illiterates. Writing poems in a computer programming language, Cramer notes, belongs to a tradition of avant-garde poetry that dates back to Oulipo's first manifesto in 1962.[72] Hence, the poem is not only a translation of Blake into an artificial programming language; it is an intertextual rewriting or transcription that both affirms Blake's poem and adds a different, rather-cynical perspective to the poem's view on the city. In short, *London.pl* measures the cries of the children "beaten enslaved fucked and exploited to death from 1792 to the present." Reading the poem is a constant switching between annotation and code:

```
# SYNOPIS and DESCRIPTION

# This Library is for redressing the gross loss to Londons Imagination of children
# beaten enslaved fucked and exploited to death from 1792 to the present.
# We see this loss in every face marked with weakness or marked with woe.
use PublicAddressSystem qw(Hampstead Westminster Lambeth Chertsey);
# PublicAddressSystem is an I/O library for the manipulation of the wheelen
# Vortex4 129db outside warning system.
#
# from Hampstead in the North, to Peckham in the South,
# from Bow in the East to Chertsey in the West.
# Find and calculate the gross lung-capacity of the children screaming from 1792 to the present
# calculate the air displacement needed to represent the public scream
# set PublicAddressSystem intance and transmit the output.
# to do this we approximate that there are 7452520 or so faces that live in the charter'd streets
of London.
# Found near where the charter'd Thames does flow.
```

The Algorithmic Gaze: *London.pl*

What distinguishes Harwood's from Blake's version of *London*, as the above excerpt shows, happens on the level of perception of the city: it replaces the perceptual gaze of the wanderer with a computational algorithmic gaze. Stephen Graham has used the term "algorithmic gaze" in relation to visual warfare technologies (e.g., drones and other tracking mechanisms), which change the cities of the Global South into battle spaces, and their citizens into potential targets.[73] Frieder Nake has suggested the term, however, as a way of expressing not the visual but rather the mental process of programming computers. In the 1960s, Nake used the high-precision drawing machine, the ZUSE Graphomat Z64, to produce his seminal prints (which were among the first computer-generated artworks to be exhibited in an arts context, including Cybernetic Serendipity in London in 1968). A main challenge for Nake and others working with computer graphics at the time was that the actual plotting of the image could take several hours. This meant that the artist's programming was as much a *mental* appropriation of the actions of the algorithms as it was a making of a drawing. In this perspective, the term is not only visual but also alludes to an antisubjective endeavor of computing a gaze. To Nake, this endeavor was an intrinsic part of "information aesthetics," as his mentor Max Bense theorized it in the 1950s (not to be confused with the current use of information aesthetics as data visualization). This "antisubjective program for aesthetics" can be seen as a reaction against Nazi Germany's strong valorization and political use of sensual perception. Information aesthetics is an objectified method of evaluating aesthetic objects.[74]

In other words, if *London.pl* bears a resemblance to an information aesthetics, it is not because it can produce poems like Blake's (i.e., translate Blake into a program) but instead because it transforms the cries of children—this aesthetic thing that affects the protagonist—into something measurable; it replaces sensual and aesthetic perception with aesthetic measure. The protagonist's personal indignation and perception of sound—the thing that remains despite the oppression of a social class; the part of life that is not capitalized—is replaced with an impersonal calculation of values (height, life expectancy, population, lung capacity, etc.) that produces a cry when the program is executed. Conversely, every cry produced by the program sets in motion not only the machine that plays the sound but also the conceptual and *mental* labor of statistics, calculation, and valorization; this is what we, the readers, sense.

As Matthew Fuller notes in his interpretation of the poem, "Screams in poetry are often representatives of an unnameable thing, a burning kernel of anguish which represents the soul and is inaccessible to language. *London.pl* by contrast … shows how much this screaming is caught up in systems of numericalisation and acceleration through operations of calculus."[75] Above all, the poem expresses a sensibility about this kind of numerical work, which is conceptual and repeatable at all times. In *London.pl*, in other words, the cries of *London* are not only the result of a particular human action at a given time but also the general operations of calculus on a fictive, imaginary module, the "PublicAddressSystem." Harwood's poem directly accounts for the calculus itself: it is a program that can run on a machine and produce the cries of a social class—not as a cry of humans, but as a fact of numerical representation and calculus. It makes the operations of algorithms and databases sensible as well as objects of thought.

London.pl is syntactically correct but incomplete and imaginary, since as pointed out by Cramer, it calls the imaginary software component PublicAdressSystem. In this sense it is only a fragment, but "exactly through this fragmentation, it gains its monstrosity," which Cramer compares to Marcel Duchamp's equally imaginary and monstrous bridal machines.[76] Yet neither Cramer nor Fuller comment on the illuminated version of *London.pl* (figure 3.7), which echoes Blake's original, illustrated poem. Of course, this visual dimension of the script is just as lost as the screams of children that are datafied and calculated by the program, since if executed, the compiler does not "see" the images in the same way that it does not "hear" the screams but only calculates the data. This loss is reminiscent of how datafication in general necessitates a loss of context in order to be executable and generate results; this loss is part of the sometimes-brutal functionality of datafication.[77]

One may speculate if, and how, this datafication and calculation is an instrument of oppression. In the final chapter of her seminal book *The Death and Life of Great American Cities* (titled *The Kind of Problem a City Is*), Jacobs suggest a direct relation between urban development strategies and the progression of science.[78] More specifically, progressions in statistics in the twentieth century made it possible to manage cities as complex organisms. As seen in the urban renewal movement, for instance, statistics formed the basis for describing the reality of working-class neighborhoods (unemployment rates, child mortality, etc.), and made way for planning the city accordingly as a homogeneous entity (with no sensitivity to the actual interactions that make up the organism).

Also at the time of Blake, calculus and the keeping of databases was an important part of the British Empire and globalized trade (including the slave trade), and a foundation of the repression described in the poem. If the culture industry of the metainterface is based on the control and capitalization of language itself as calculus and database, as described in chapter 2 (with all the implications this has for creativity and the formation of subjectivity), it is also a form of spatial control—an urban setting where screams and other exclamations of the human soul, along with (and on the same level as) everything else, become objects of datafication, calculus, and production.

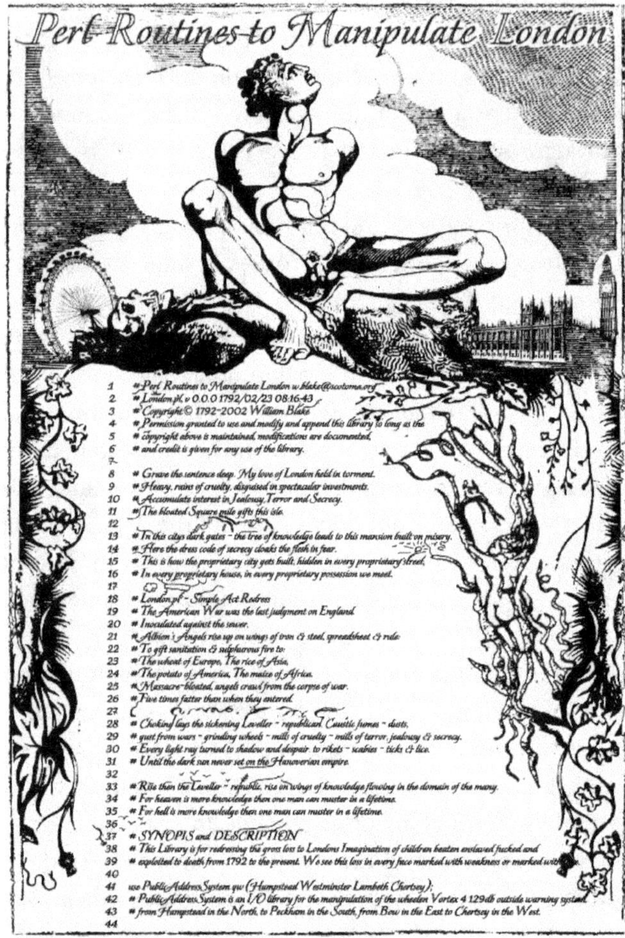

Figure 3.7

Perl Routines to Manipulate London—London.pl (2002) by Graham Harwood. *London.pl* is an operational version of William Blake's famous poem. The executed perl program code (though not fully functioning) measures the cries of the children "beaten enslaved fucked and exploited to death from 1792 to the present." The illuminated version of *London.pl* also echoes Blake's originally illustrated poem. Illuminated version of poem by Yoha.

The Open and Free

If the city is perceived as a text that can be writerly, then, as argued above, the urban metainterface operates in several writerly ways. The urban metainterface addresses an optical dimension that is parallel to a calculative and computational dimension. As examples, *Hole in Space*, *Body Movies*, and *Serendipitor* all demonstrate that both cameras and screens along with the instructions can be used to open up the city, and reconfigure how it is perceived, lived, and conceptualized. Yet *Faceless* and *London.pl* also point to a different writerly dimension: the interface itself. This is central to the understanding and critique of the urban metainterface. As much as urban cameras, screens, data, and algorithms, applied to anything from CCTVs to Google maps and other apps, may open up new configurations of meaning, giving way to alternative spatial practices, personal relations, and representations as well as behaviors that are not always anticipated, the interface itself holds within it a particular optical and algorithmic gaze. To open this gaze involves a different kind of openness that is not only related to the open processes of signification. As is evident in both *Faceless* and *London.pl*, the open process of signification is driven by a profound interest in freedom—in setting the human free, and giving the human soul a voice through the technological means of production that conditions it. This freedom is significant and important, but how can it be conceptualized?

As Kit Galloway remarks in an interview with Annmarie Chandler, "The age of the avantpreneurs is over. We now return to an age of the entrepreneur."[79] This implies that the artist, who was once in the streets exploring the intrinsic human and social qualities of new technologies for communication, is now back in the gallery; but perhaps more interesting, it also implies that being an avantpreneur, exploring how technologies may potentially open up space for social formation, as Galloway and Rabinowitz have investigated extensively throughout their careers, is now considered entrepreneurship. To create a hole in space, dérive, chance encounter, and so on, are practices that are remarkably indistinguishable from social smartphone applications that provide an easy interface to potential social encounters with strangers (e.g., dating apps), constant flows of location tagged images (e.g., Instagram), and so forth. "Opening the city" has in this sense become a mantra, not only for the artistic use of information technologies, but for urban entrepreneurship and innovation, too. In this sense, the ideals of openness also seem far from the ideals of freedom that are present in, for instance, *Faceless* and *London.pl*. Or rather, the open and free express two writerly approaches to the urban metainterface that are similar and yet fundamentally different.

The difference is well known within software culture, as a distinction between open-source and free software. The ideal of opening the city as well as other kinds of openness in governance and elsewhere (open data, open education, open access publishing, etc.) should be seen in light of the success of open and agile software development. As has also been noted by Nathaniel Tkacz, openness is one of those terms that avoid criticism by being put forward negatively (i.e., "not closed"). This lack of criticism gives it an imaginary air that makes it appealing to visions of future societies, such as Michael Hardt and Antonio Negri's use the term "open-source society" in their depiction of a free society (or a democracy of the multitude), thereby explicitly relating the conception of society to the conception of software.[80]

To make up for the conceptual shortcomings of openness, Tkacz conducted a thorough study of its philosophical origin. In other words, rather than using openness to look forward, as is the common practice, Tkacz looks backward into the history of openness to gain a more critical understanding of the concept. Its application today is inspired by the success of open-source software development, but naturally this culture has found its ideals somewhere. Tkacz specifically discusses Karl Popper, who during his time thought it particularly necessary to critique intentions to build ideal societies (such as Nazism or Communism). Popper's idealization of openness also corresponds well with Friedrich Hayek's promotion of neoliberalism in the 1940s. The direction of a society is beyond any individual or group's knowledge, and only open competition between ideas and practices will provide the agility that ensures liberty and prevents totalitarianism.

Hayek and Popper's ideas resonate well, as Tkacz points out, with the perception of openness within software culture, and how this stands in contrast to "free" software development. The promotion of both open and free software appears as a response to proprietary and closed software formats, but follows two competing routes. One route is Richard Stallman's ideas of a GNU General Public License (GPL) as a general software license associated with the free software movement: all software that includes work issued under the GNU GPL must inherit the same license. The other is Eric Raymond's advocacy for an Open Source Initiative that explicitly sees Stallman's political projects as a hindrance to technical development. According to Raymond, competition through commodification is a much more efficient way of developing software that meets the desires of the users. To promote competition, licenses such as the Mozilla Public License therefore insists on the open availability of the source code, but also allows for proprietary modules. The possibility of capitalizing on one's work ideally better supports future progress. To the Open Source Initiative, commodification is the ideal "bazaar" of software development, contrary to the "cathedral" (as Raymond famously titled his influential *The Cathedral and the Bazaar*).[81]

In other words, contrary to the free software movement that rests on an ideal freedom of all humans and a world that is a shared resource (e.g., it insists on the collective ownership of the technical infrastructures), the Open Source Initiative is much more aligned with the neoliberal openness envisioned by Hayek. As the avantpreneur becomes an entrepreneur, it seems as if it is Raymond's neoliberal and market-oriented vision of an open society that is put into play, and not Stallman's notion of free, which is more socially aware, and less focused on capital and personal gain (free is also often referred to as "libre" to highlight this difference).

Instagram, TripAdvisor, Airbnb, Uber, and many other urban metainterfaces today correspond well with the kinds of neoliberal thinking that can be associated with openness. They thrive on open access not only to the city itself and its resources but also the city's hidden layers of data and the right to combine data to generate new uses of the city. Rather than claiming that *Serendipitor* and *Hole in Space* are neoliberal in their insistence on opening up the city on its visual and organizational levels, and that *Faceless* and *London.pl* are free, though, the above examples demonstrate an aesthetic and not an ideological difference. If *Serendipitor* and *Hole in Space* open up the processes of signification in the city, then *Faceless* and *London.pl* distinguish themselves by setting the machinery that produces meaning free. In this, the software license does not seem to matter; instead, it is the ability to access and reflect an inner tendency within urban surveillance and numerical representation that matters, and how the aesthetics reflect the operational dimensions of this. In other words, the metainterface is not only a means to access the city but also distributes meaning in particular ways that can be reflected as a tendency.

Designing Open Urban Metainterfaces

Strategies of openness not only function as a new and networked business, where the platforms of the metainterface industry are extended into urban space but often specifically seek to provide sustainable urban innovation. For instance, the Dutch urban design research project, "Hackable City" seeks to "explore the opportunities as well as challenges of the rise of new media technologies for an open, democratic process of collaborative citymaking."[82] Strategies of hacking, bricolage (learning by doing), and collaboration that reverse the relation between expert city planner and the amateur citizen become a way of opening up space for new, socially and environmentally sustainable urban designs.

The Hackable City indeed seems be an appropriate term for a description of a city where the implementation and use of pervasive technologies are not brought about by science labs but instead a cultural logic of openness and access to data and code. As noted by Adam Greenfield, too, the self-proclaimed prophets of ubiquitous technologies (Mark Weiser, Don Norman, and others) are not the drivers of innovation: "There seems to be little interest in the various 'digital home' scenarios, even among the cohort of consumers who could afford such things and have been comparatively enthusiastic about high-end home theater."[83] Rather, innovation occurs when, for instance, metadata from a photo of a tool for loan (such as its location) is combined with Google Maps to produce a marketplace, where people can search and identify items in a community marketplace. If only the infrastructures and flows of data were open, they would inevitably be mashed-up by third-party applications. This is the cultural logic of net culture, which is far more flexible than foreseen by the lab prophets who seek to educate users of a technology application's desirability "until they manage to work up the appropriate level of enthusiasm," as Greenfield observes sarcastically.[84]

Frequently, the innovation of metainterfaces that generate new value in the city (financial as well as social or environmental) rely on open access to data. They rely on the free right to combine the data of users' locations, movements, photos, music, and so forth, to generate new services. Tim O'Reilly perhaps best expresses this successful partnership between entrepreneurship and open data. In a discussion on "open data for open land," he writes, "This policy was based on lessons from previous government open data success stories, such as weather data and GPS, which form the basis for countless commercial services that we take for granted today and that deliver enormous value to society." As a rationale, he refers to GovLab's Open Data 500 project as an "impressive list of companies reliant on open government data."[85]

Arguably, traffic monitoring systems, apps for sharing and urban activities, and many more that "hack" the city, do often succeed in building new urban communities, and do suggest sustainable alternatives to, say, the ownership of consumer culture, traffic congestion, public health, and so forth. They also present new spatial practices, however, and sometimes these practices have unforeseen consequences and lead to new types of conflicts. To urban governors as well as many drivers, a traffic monitoring system that combines information from parking lots with a route planner may be extremely useful, but it also bypasses the general discussions of traffic and political solutions that would favor bicycles, public transportation, or pedestrians instead of private cars. The open designs, in other words, are not only interfaces that reconfigure the use of urban space; they themselves contain their own grammar or inner tendency, too. The urban metainterfaces have their own unconscious, hidden scripting of space that produces new types of anxieties and mass ornamentation. This also creates new needs and desires to see, stage, and critically reflect this new mode of production (as demonstrated in the works of Luksch and Harwood, for example).

To uncover such alternative strategies for using the urban metainterface as an expressive language, the final part of this chapter discusses how interface and network culture simultaneously may employ alternative strategies for "free" expression.

Designing Free and Critical Urban Metainterfaces

The aims of this chapter have not been to oppose open data politics, or the usability of apps such as Endomondo, Airbnb, and others, but rather to challenge their inner logics and grammar. The instrumentalized dream of openness to social behaviors and patterns is benevolent by nature, yet may potentially disarm any negotiation. What is open access to urban cameras, screens, data, and algorithms, other than the right for anyone to apply and even harness their inner logic?

In his book *Two Bits*, the US anthropologist Christopher M. Kelty demonstrates how free software culture's obsession with openness in technical infrastructures is not only a question of technical efficiency or solutionism but also politics along with critical discussion within software culture. This critique, as Tkacz has also pointed out, has been the impetus for forming alternative designs and ways of licensing them. In Kelty's terms, free software represent a recursive public, where people care deeply about the system they inhabit. He defines a recursive public as "a public that is vitally concerned with the material and practical maintenance and modification of the technical, legal, practical, and conceptual means of its own existence as a public; it is a collective independent of other forms of constituted power and is capable of speaking to existing forms of power through the production of actually existing alternatives."[86]

Free software literally includes a critical public as part of its production and design. When it comes to the design of software and networks that have real and political consequences, such as the Internet, this critical public is essential. In the free software movement, there are continuing opportunities to criticize, revise, and make alternatives, since the design process is not hidden behind a closed proprietary wall. If the Internet changes the conditions of social organization, then "Free Software is, arguably, the best example of an attempt to make this transformation public, to ensure that it uses the advantages of adaptability as critique to counter the power of planning as control," as Kelty writes. Recursiveness is a way to ensure a process that is critical and pursues alternatives. "It is a bit like Kant's version of enlightenment: insofar as geeks speak (or hack) as scholars, in a public realm, they have a right to propose criticisms and changes of any sort."[87]

In relation to the urban metainterface as well as the ways cameras, screens, data capture, and algorithms are applied in cites, the recursive public of free software lends itself as a usable model for urban interface design.[88] The important difference with neoliberal thinking that is often applied by the metainterface industry is that this would be a critical process that includes politics and does not aim to just make "smart" solutions in the city. It would allow for a writerly urban metainterface as well as a dialectics between forms of criticism and the urban text, as suggested by Benjamin in *One-Way Street*. Almost echoing Benjamin, Kelty finishes his book by asserting, "There is no doubt room for critique—and many scholars will demand it—but scholarly critique will have to learn how to sit, easily or uneasily, with Free Software *as critique*." Such critique is highly necessary, as Kelty also argues in relation to social software, where limited conceptions of openness potentially take over. Social software platforms may appear collaborative to some degree, but lack the recursive public commitment, since they are not "interested in allowing strangers to participate in, modulate, or modify the system as such" but mainly "in allowing users to become consumers in more and more sophisticated ways."[89] Consequently, there is an immediate risk of a metainterface-industrial takeover of the urban.

Any instrumentalization of openness inherently implies closures. The closure of an open urban metainterface industry seems particularly dangerous in that it is highly operational, but potentially rules out any sense of conflict and negotiation that relates to moral as well as formal visions of a society. It is a potential strategy that rules out urban politics to the benefit of slick participation, "smartness," and efficiency. There is, in other words, an enormous difference between the application of open source to "open data" (as envisioned by O'Reilly and others), on the one hand, and potential applications of free software to data, on the other. What might the latter look like? What does an alternative design approach look like? Visions from architecture may prove helpful.

In an investigation of "the city as an open system," the urban theorist Richard Sennett reinvigorates his (and many other urban theorists, activists, and architects) mentor, Jane Jacobs. To Jacobs, the idea of an open city includes ad hoc adaptations and diverse use of public space (such as having shops, garages, health care, and so on, in the same neighborhoods). It is a city where everyday use leads to mutation and chance variation. Though Jacobs' ideal city is essentially an open city, Sennett also detects "glints of something lurking beneath" the stark contrast Jacobs draws between the homogeneity of closed systems, on the one side, and the diversity of open systems, on the other. Jacobs is an "anarchist of a peculiar sort"—a much more conservative anarchist, Sennett notes. It is the urban cultures that have "taken root" that are agile, and able to produce as well as absorb chance and change: "It is why Naples, Cairo, or New York's Lower East Side, though resource poor, still 'work' in the sense that people care deeply about them."[90] For Jacobs, the question is, Is it possible to control, maintain, or in other ways plan for this open use of the city?

Sennett contends that the neoliberal open development of cities results in cities that "decay much more quickly than urban fabric inherited from the past." The constant change and mutation of the city results in "brittle cities" that—in their opposition to the closed system—serve the interests of a few brute enterprises rather than the different voices of a social system. Neoliberalism merely speaks the language of freedom "whilst manipulating closed bureaucratic systems for private gain."[91] Consequently, he argues for an alternative way—a different urban social system with a different kind of openness than the private enterprise, through what he labels passage territories, incomplete forms, and urban narratives.

Hence, Sennett suggests a competing formal version of open urban design—one that is not just focused on open competition as a way to obtain social sustainability but rather on the rooted community's own ability to handle change.[92] Sennett's objective is to speculate on how architects can design cities that are open to this process of taking root, allowing an experience of time that enables citizens to handle mutations and chance variation, and also conflicts. This is a physical experience of democracy and participation, too. Sennett's perception of a socially rooted openness corresponds better with a "free" openness and the recursive public Kelty identifies in cultures of free software. In other words, in opposition to a neoliberal urban development (which he associates with the transparency of glass windows and ongoing competition between urban elements, in which only the fittest and most competitive will survive), Sennett favors the structures that are open to internal revision according to a social moral.

As an alternative way to design for such habitation, and as working principles that challenge neoliberal urban design, Sennett suggests "porosity of territory, narrative indeterminacy and incomplete form."[93] Openness is not a question of visual transparency, and is not represented by the contemporary (i.e., around 2006) architectural use of glass facades; rather, openness should be seen as a "passage territory." He exemplifies this with the medieval wall. Unlike the conventional conception, walls can function like "membranes." The medieval wall not only forms the boundaries of a city but also attracts unregulated development, such as black markets, and more generally "heretics, foreign exiles, and other misfits."[94] The functions of the wall also highlight how incomplete forms work. Buildings are only complete when situated in a context with other buildings—not as self-referential objects. Hence architecture ought to be engineered fragmentally, and leaving subsequent conflicting practices open also creates a narrative of development. This is the kind of "Darwinist" evolution of cities, envisioned by Jacobs, that roots the community, makes it care, and enables it to deal with mutation and change, and thus also is a potential basis for a writerly semiotics (as outlined by Barthes and Benjamin).

A passage territory of an urban metainterface—which does not simply strive to apply the interface but also open it up to the kinds of incompleteness, conflicts, negotiations, and narratives that root it in a community—essentially replaces the "open" with the "free." It is a kind of interface that—like, for instance, *Spoken Streets, Faceless*, or *London.pl*—does not claim an immanent truth but instead is free to chance variation and change, and is expressive rather than indicative. We need expressive interfaces in order to understand and deal with the neoliberal use of "open data"—that is, we need to question and challenge the strategy as an ideological construct and self-fulfilling prophecy of an open society.

The question of free interface design that looks beyond the question of license agreements, and applies the notion of tendency as a critical design principle—tendency as an expressive power, and not only an analytic tool—will be further elaborated on in the final chapter. The next chapter will pursue the question of architectures as material constructs that have much more than mere functional consequences. This issue seems particularly relevant in relation to so-called cloud computing, which is the ruling technological infrastructure for the metainterface industry.

4 The Cloud Interface: Experiences of a Metainterface World

Super Mario Clouds

"The cloud" is a signifier for the computer networks of the metainterface industry. It was first mentioned in relation to the Internet in a 1996 MIT research paper, but was mainly popularized by Google CEO Eric Schmidt in 2006 as a description of strategies (already deployed not only by Google but also Yahoo!, eBay, and Amazon) to keep the corporation's data services and architecture on servers.[1] Hence the cloud is both a canny metaphor, business plan, and somewhat loosely confined technological paradigm. Technically, the cloud relies on so-called NoSQL databases (as opposed to the relational SQL database), which are massively scalable and can handle the extreme workloads that characterize cloud computing.[2] Functionally, it signifies the absence of hardware storage and processing power on the user's device. These technical infrastructures are instead located elsewhere (in the cloud), but can be accessed through the Internet. This allows for slim design, mobility, and flexibility, but as discussed earlier in this book, it also forms the basis for a new consumer industry: a metainterface industry based on the production and consumption of culture, and that reorganizes cultural practices as well as urban space.

Both the functionality and technical specifications of cloud computing have spawned many critical conversations about surveillance, censorship, the restriction of access, the legal rights of files and data, and more. For instance, the research-focused design studio Metahaven (founded by Amsterdam-based Vinca Kruk and Daniel van der Velden) defines the cloud as "the informational equivalent to the container terminal," with a "higher degree of standardization and scalability than most earlier forms of networked information and communication technology."[3] In its look at censorship, surveillance, and the controversies between the National Security Agency and Edward Snowden, Metahaven points to how the cloud can be seen as a new business concept in line with commercial and national (and specifically United States') interests—rather than the user's interests. Above all, the cloud therefore functions as an "architecture of power" in which the "early internet is eclipsed."[4]

Consequently, cloud computing can be seen in a historical perspective, as the result of a development where technological models are refined to incorporate aspects of net culture into new business models. This can even be viewed through the way aspects of net culture reflect back at us today, making us see what has disappeared, and how it has become incorporated into the cloud of the interface industry. Facebook, for example, has become a template-based, monitored version of what Usenet, Geocities, and other phenomena for personal communication and social interaction were in the 1980s and 1990s. In other words, the early Internet of personal home pages has been taken over by cloud services and corporations that combine the storage, distribution, and software services of the interface industry within a "single black box."[5]

The main objective of this chapter is to explore how this commercial and national standardization influences the experience of the interface, and how this cloud interface affects sense perception on a more general level. What tendency lies within the production of cloud interface experiences? According to Metahaven, the users' experiences of cloud interfaces consist of a "sense of abstraction from prior experiences; in the cloud, the user no longer needs to understand how a software program works or where his or her data really is."[6] The New York–based artist Cory Arcangel has aesthetically examined this sense of abstraction from software and data, even in work that predates the cloud. Arcangel is famous for modifying and appropriating materials and standards from popular interface technologies, such as websites, games, image editing, terminal scripting, and more. His work can be characterized as an expression of a do-it-yourself network culture, but also an arts practice that on several occasions, has been brought into established arts institutions as "postconceptual" and "post-Internet" art that reflects the Internet's conditioning of aesthetics and culture. One might say that the arts institution has recognized the tendency operating within net art, and that Arcangel has been extremely skillful at working with interface effects in humorous and reflective ways. In his tools, performances, and images, many people will recognize a parody of their everyday experiences with the networked computer: elements and routines that are not always consciously reflected are suddenly repeated, looped, or in other ways highlighted and brought forward.

In 2002, Arcangel made a modification of the video game *Super Mario Bros.* for the Nintendo Entertainment System. The game, from 1985, is something that most people of Arcangel's generation recognize and have played in their childhood. Arcangel modified the cartridge of the original game in such a way that Mario, every item in his Mushroom kingdom, and the player statuses (time, lives, points, etc.) have disappeared.

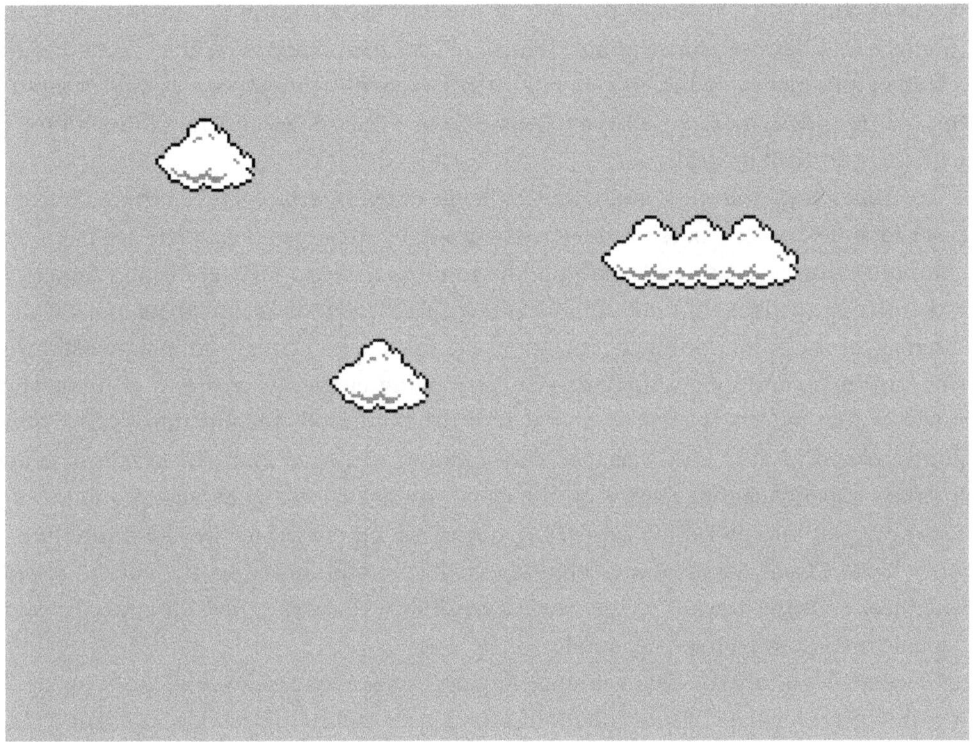

Figure 4.1
Cory Arcangel, *Super Mario Clouds*, 2002. Handmade hacked Super Mario Brothers cartridge and Nintendo NES video game system. Edition no. 2/5. Whitney Museum of American Art, New York; Purchase with funds from the Painting and Sculpture Committee (2005.10). © Cory Arcangel. Courtesy of the artist.

What is left is only the game's background: the blue sky and white clouds floating across the screen. With no interaction, no game, and no rules, the viewer is confronted with the visual backdrop of the interface: its calm and comforting two-dimensional, minimalistic milieu where nothing seems to happen.[7]

Super Mario Clouds (figure 4.1) can be seen as a contemplative interface that imitates the experience of lying on one's back and looking at the passing clouds on a sunny day. This is probably how it was intended. Yet the removed elements and scripts also leave the user with a sense of a missing interface—a technological void that somehow becomes an aesthetic experience of pleasure when looking into the clouds. It is a simultaneous experience of meaning and absence. *Super Mario Clouds* was developed long before cloud computing, but it contains much more than a metaphoric reference to cloud computing: it stages the lack of interface as an aesthetic experience of an interface that has been blocked out. Hence in a contemporary view, the spectator sees a kind of minimalism (which is generally associated with contemporary cloud computing interface design) as an abstract space whose dimensions and functionalities are extremely difficult to work out.

In other words, the epistemological challenge of the interface—the abstraction from prior experiences—has been studied aesthetically by Arcangel, but in the age of cloud computing this sense of abstraction is the result of a technological business strategy rather than an artistic exploration. By making *Super Mario Clouds*, Arcangel looks at the interface of the Nintendo game, including the rules, algorithms, and material storage (the cartridge), and by manipulating and removing elements, creates a minimalistic interface that reflects the game interfaces of his childhood. Yet this minimalism also points toward what in 2002 was the future of interfaces. *Super Mario Clouds* displays an interface experience that reflects on the cloud interface: looking at *Super Mario Clouds* today, we can interpret how it reflects a way of seeing related to the cloud interface.[8] *Super Mario Clouds* demonstrates how aspects of net culture reflect the current cloud interface, pointing toward elements of accessibility that have now disappeared and become incorporated into the cloud.

If interface culture has changed since Arcangel conceived *Super Mario Clouds* in 2002, it is not only a change of user functions afforded by technological developments. The cloud computing of the metainterface industry, along with its control of consumption and production, and de- and reterritorialization of urban space, is also an epistemological challenge, or a question of handling a "sense of abstraction of prior experiences," as Metahaven argues. As an epistemological challenge, it also exceeds the interface. This chapter thus addresses the absent materiality and architecture that generally characterizes cloud computing as a question of how one perceives reality—ranging from the interface itself to seeing the world as an interface. More specifically, it addresses artworks that deal with this kind of production of knowledge in two interrelated ways: the (absent) architecture and infrastructure of the metainterface, and how this relates to the perception of reality, such as the war on terror as well as the signs of environmental and climate change. It addresses works, which in their realism reveal a tendency in production. In other words, the production of the works more generally relates to how and what cloud computing conceals, and how this concealment makes the user perceive the world differently.

The Architecture of the Vernacular Web: *Summer*

Similarly to Metahaven's description of how the early Internet is eclipsed and black boxed, Olia Lialina, the German-based Russian net artist and professor at Merz Akademie in Stuttgart, and her collaborator Dragon Espenschied, a musician and media artist, describe a cultural shift from the development of net culture. Looking at how amateur websites have been integrated and absorbed by Web 2.0 and cloud computing, they argue that the Internet has moved away from a "vernacular web," where "users could easily write the code for their own web pages," toward a web where the users are placed "in front of dumb terminals, feeding centralized databases."[9] In many ways, the practices of the vernacular web literally built the Internet. "People felt it was their personal responsibility to configure the environment and build the infrastructure." Gradually in the late 1990s, however, search engines, portals, and catalogs "took over the linking responsibilities."[10] Today, the links and networks between pages and people—the architecture of the web that to a large degree was designed and curated on independent individual home pages—have become functions and services, such as "friends" and "shares" on Facebook, or "followers" and "retweets" on Twitter. The network, originally curated, cared for, and constructed by independent users, has increasingly been taken over by cloud platforms, and the link has gradually shifted from being a sign of individual reference and recognition, where the user links to something in order to create coherence, to a central infrastructural function where likes and shares generate data for behavioral profiles, and where this is used to autogenerate "recommendations." Whereas the early 2000s led to a proliferation of hypertextual linking—coauthored, multisequential, and multicursal writing (for example, on blogs and wikis)—these hypertextual trademarks increasingly have become part of the designed functionality of Web 2.0 platforms.

An illustration of this is Facebook, where the functional hypertext activity of linking is staged and performed as sharing, liking, befriending, and other personal and affective actions. This staging of hypertext as a social infrastructure produces behavioral data and profiling that Facebook uses to generate the multicursal, coauthored text on its users' News Feeds. In other words, linking and hypertext, which were once seen as a rhizomatic and associative writing strategies, have become recentralized and monopolized by cloud platforms. As discussed in chapter 2, and pointed out by John Cayley and Daniel Howe's work on Google (*The Readers Project*), such strategies of the monopolization of semantic acts are not only part of social media but also the metainterface industry more generally. According to Lialina and Espenschied, though, there are still traces of the older and more open structures of the premillennium Internet. They have

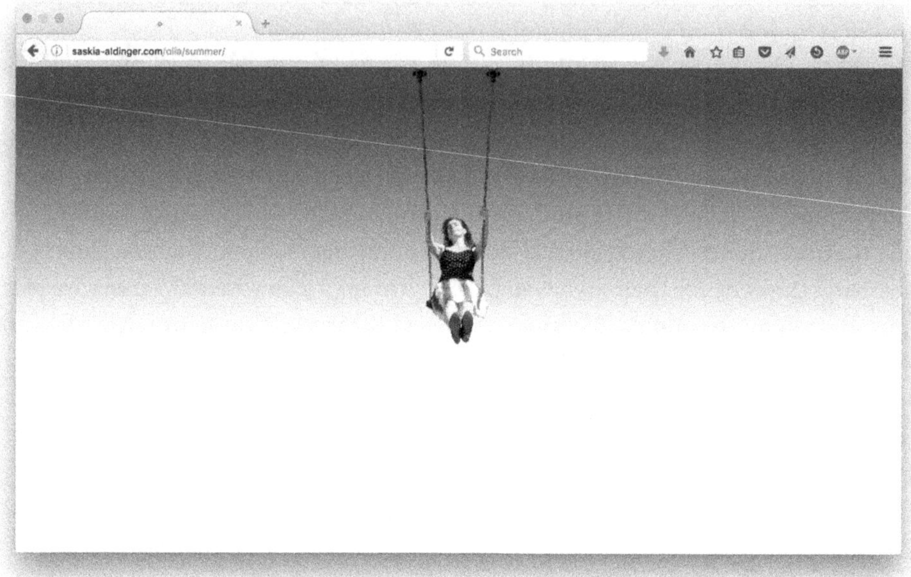

Figure 4.2
Summer (2013) by Olia Lialina. An animated GIF of Lialina swinging. The animation is dependent on personal as much as technical relations. The animated GIF has 21 frames, each of them are stored at a different server, and hosted by people in Lialina's personal network of friends. Each server needs to be accessed to produce the effect of Lialina swinging. Courtesy of Olia Lialina. Still (screen capture) from animation.

explored this in their artistic research on the vernacular web and digital folklore, where they have been collecting and documenting animated GIFs.

Unlike the compression format JPG, the GIF image format can be used to display and loop animations. This made the format popular in the early days of the World Wide Web, but the animated GIFs have recently gained increased popularity via meme culture (with meme culture's usual sense of parody and retrospect). In 2013, Lialina made the net art piece *Summer* (figure 4.2), which shows an animated GIF of her dressed in summer clothes and swinging in bright sunshine on a playground swing that appears as if attached to the top of the browser's address bar.[11] The animated GIF has twenty-one frames, and though the visual impression of the film, creates a light summerly feeling, it plays back somewhat irregularly. The visibility of the animation's composition—its techniques of looping and montage—is particular for the animated GIF. It is an expression that generally does not try to hide its materiality. In *Summer*, however, the montage is sometimes especially visible—as if some of the images in the animation are missing. This is because the images in the GIF are not stored on one web server but instead on twenty-one different web servers belonging to people in Lialina's personal network of artists. Each server needs to be addressed and accessed to produce the effect of Lialina swinging from the browser's location bar.

Watching *Summer*—its stuttering and fragile playback as well as the constantly shifting URL's in the browser's location bar—the viewer is constantly reminded of the networked montage character of the piece. Lialina expresses it herself: "I like to swing on the location bar of the browser, and I like to know that the speed of swinging depends on the connection speed, and that you can't watch this GIF offline."[12] To put it differently, the architecture of the web that sometimes keeps us waiting for images to load is not just a question of functionality; rather, it is a sign of the vernacular web: people, servers, and a built environment. This, of course, stands in sharp contrast to the smoothness of the network that meets the user in the age of cloud computing, where the robust network—with high-speed data transfer, efficient data buffering, and high-performance storage management—is an essential quality. *Summer* points to how the architecture has become a trivial circumstance, "out there," "in the cloud," and how the user is constantly prompted to leave the question of architecture to the specialized enterprise (the smooth operators).

Summer echoes one of the defining pieces of net art: *Refresh* from 1996. *Refresh* was organized by the Russian net artist Alexei Shulgin as a collaborative work, where a number of net artists were invited to show a small piece of their work on their web server for ten seconds, and then quickly forward the user to the next piece, using the <refresh> HTML tag that forces the web page to reload from a different URL. *Refresh* consequently is a net art piece of its own as well a network of net artists and net artworks.[13] In this way, it is a perfect example of how users—in this case, net artists—were "literally building the Internet."[14] The piece is almost gone today because the original servers and/or addresses have disappeared. Lialina's own former work *Agatha Appears* from 1997, which is also spread out on different servers, and uses the shifting URLs as parts of its narrative, comes into mind when watching *Summer*.[15] Hence, *Summer* brings back memories of not only summer days but former anticipations, too: the imaginaries and dreams of the creative and social potentials of a decentralized, collaborative World Wide Web. It reminds us of the power of linking and hypertext—including the physical and social architecture of servers in a network—as tools to build and explore networks, rather than just a smooth interaction on a smartphone or another cloud platform. A newer artistic counterexample to streaming platforms is *The Pirate Cinema* (figure 4.3), a project that explores the torrent protocol of peer-to-peer networks along

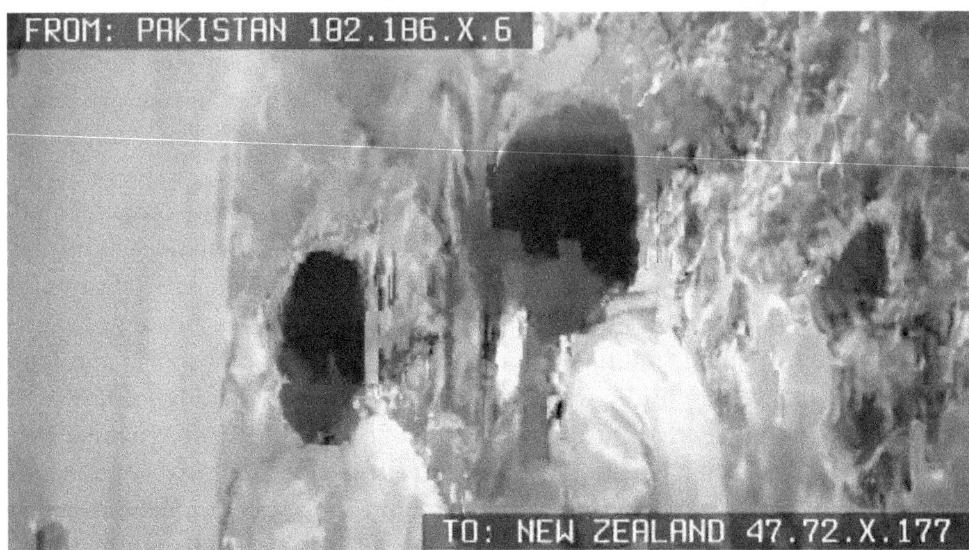

Figure 4.3
Pirate Cinema (2012–2015) by Nicolas Maigret. *The Pirate Cinema* is a realtime display of BitTorrent files from the top 100 videos on the peer-to-peer file sharing system, The Pirate Bay. This live exchange of files in transit between recipients and emitters, unknown to each other, form a continuous and fragmentary audio-visual mash-up. Courtesy of Nicolas Maigret. Still (screen capture) from online installation.

with the potential for viral transmission and alternative social models. It downloads popular torrents, and shows them as a collage in a nonlinear fashion and mixing different films. In this way, it makes "the hidden activity and geography of peer-to-peer file sharing visible," as described by Geoff Cox, and depicts how there is a topology behind the real-time and global exchange of files and data: although the streams of the cloud appear out of time and place, they point to the microtemporal and infrastructural processes that easily escape perception in real-time streaming.[16]

Summer's return to the net art aesthetics of the mid-1990s highlights how a generation of net artists were prefiguring a web of constant flows and streams, but also how this eventually turned into multicursal streams of newsfeeds and the like. It creates an awareness of how the amateur developer of the vernacular web has been incorporated into the templates and platforms of the interface industry—turning the affective actions of liking and sharing by building network architectures into passive consumption (clicking), and the passive consumption of reading, listening, and watching into the active production of enterprise data centers and other architectures, hidden away and physically protected from the threat of human intervention. While the coproducing user has been celebrated, such as the person of the year of *Time* magazine at the end of 2006, "this grand 'comeback' of the user ... illustrated just how vast the gap between users and their computers had grown," as Lialina and Espenschied write. *Summer* points to how the smoothness of images becomes something expected when the vernacular architectures are taken over by enterprises—something that makes the image appear close, but nevertheless opens a gap between the user and architecture. This gap functions in different ways.

First, it creates an alienating smoothness. The stuttering and demonstrative signs of the shifting URL locations in Lialina's *Summer* are gone in cloud computing. In cloud computing, it is impossible to determine where the stream comes from, and the user is not made conscious of the shifting locations of their data or the software platform they are using. Instead, the montage character so present in *Summer* is turned into continuous streams and a simulated seamlessness where the user loses touch of the underlying network. The combination of data packets is invisible to the user, and if one experiences stuttering when streaming, it is seen as an error without further explanation and not a desired aesthetic network effect.

Second, this smoothness operates within other media, too. Within music, for instance, the listener experiences an abstraction of the sound in relation to its source. Jeremy Morris has argued that music listening with cloud and streaming services becomes ubiquitous, and is accompanied by a "certain 'sourcelessness' to it." He quotes Anahid Kassabian's characterization of cloud-based music as something that "comes from everywhere and nowhere. Its projection looks to erase its production as much as possible, posing instead as a quality of the environment."[17] Listening (as well as watching, gaming, clicking, tweeting, etc.) becomes casual, mobile, and detached.

Third, it creates a network architecture that obstructs human intervention. In contrast to the vernacular web, the cloud network has become a streaming platform that hides its networked infrastructure. By doing this, it turns into a virtual media channel that simulates older media platforms like TV or radio, where the streaming architecture impedes sharing, remixing, or unauthorized access. This has, for example, significance for music culture because the user has less control over the collection of music: "While CDs and tapes only required one other device for playback (*i.e.*, a CD player or tape player), music in the cloud is contingent on a network of technologies, devices and connections," as Morris notices.[18] This integration of technologies and devices obviously also relates to the interface industry described in chapter 2.

Fourth, it creates inequality between users and platform providers. While users might not know where their streaming comes from, the streaming platforms know exactly where and who receives the stream. This information is generally used to abide by geographic copyright distinctions and control the range of content provided on the platform, and also create predictive algorithms of taste and preferences that leave the users with the content they are most likely to like. In other words, the smoothness of the platforms erodes the more equal relationship of the earlier net culture between users and content providers—a relation that is now condemned as "piracy."

In short, the cloud interface creates a sense of abstraction from prior experiences, a sense of media absence where the user experiences everything as ready-at-hand real-time streams and no longer needs to care for the media, or where the sound, the images, and so on, are located, and how they are processed, stored, and accessed. Media and mediation is already an example of abstraction, but the cloud interface's "meta-abstraction" produces a general abstraction and absence on a global scale. With Google Drive, iCloud, and numerous other services, the cloud has become an industry that produces an airy symbolism. This symbolism is a sign of not only new and smooth functionality but also absence, and a meta-abstraction that has become generalized (all over the interface and in every interface) and globalized (all over the planet). This stands in contrast to the dreams of a decentralized vernacular web where people cared for its infrastructure, where "architecture" is not only an infrastructure for content dissemination but also seen as a recursive public that forms a community and takes root, as discussed with the concepts of Christopher Kelty and Richard Sennett in chapter 3.

Although the cloud is primarily experienced as a smooth abstraction, from everywhere and nowhere, that is internalized by its everyday users, *Summer* indirectly suggests the architecture of the network today. What is this architecture of the cloud as a thing that is cared for, and a thing that can be judged aesthetically or criticized for the life it offers its users ("in front of dumb terminals, feeding centralized databases")?[19] The following presents three works that each in their own way exemplifies how cloud computing displaces its architecture by building on principles of virtualization, and how this is a displacement of time and place that conceals real, material, and political conflicts behind a veil of rhetoric and smoothness. All together, the examples point toward how the cloud functions as a phantasmagoria, or shadow play, of contemporary globalization.

Clouded Architecture and Metainterfaces: *Internet Machine*

If the clouds of the interface industry stand between the user and architecture, this by no means implies that the cloud has no architecture. The architecture merely looks different. Monstrous data centers and other architectures that are hidden away, protected

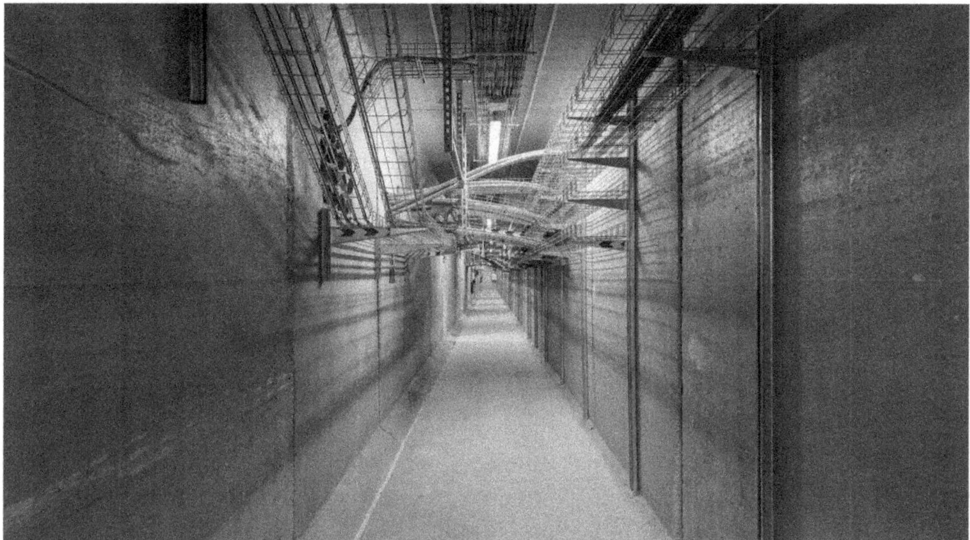

Figure 4.4
Internet Machine (2014) by Timo Arnall. The footage shows the global broadband and telecommunication company Telefónica's large data center in Alcalá, Spain. The work draws attention to the material, and also geopolitical, aspects of data-transactions in cloud computing. Courtesy of Timo Arnall. Photo by Timo Arnall.

from the threat of human intervention, have replaced a network of personal web servers. The architecture has become a cold object that is cared for by "other people" and not a matter of concern (a "thing") for the user, as in the vernacular web.[20]

Several critical and artistic projects investigate this concealed architecture of the cloud—the cable routes, routers, data centers, and other logistic elements. One example is the visual artist and designer Timo Arnall's multiscreen film *Internet Machine* from 2014 (figure 4.4). The film shows the global broadband and telecommunication company Telefónica's large data center in Alcalá, Spain. Arnall claims that the film was made "to look beyond the childish myth of 'the cloud,' to investigate what the infrastructures of the Internet actually look like."[21] The slow camera movements and high-definition multiscreens of the film, however, also produce a kind of noir aesthetization of the cloud. The enormous high-tech cathedral may even be seen as a fetishization of the cloud's architecture. Arnall's work produces a place for the abstract cloud and draws attention to its geopolitical aspects; it produces a place for the carrier of data transactions, telecommunication, and more that we commonly refer to as central to globalization. The place for globalization appears as a place out of human reach, almost sacred.

But as an ordinary user, one may ask how seeing a film of a big server farm in Spain relates to the experience of a metainterface.

On a more fundamental level, the movie's illustration of the cloud's architecture reverses a function that is common not only to globalization but also computer science more generally: virtualization. The United States–based researcher and network engineer Tung-Hui Hu elaborates on this aspect of the cloud in his book *A Prehistory of the Cloud*. Hu explains how the cloud emerged as a symbol from ways of drawing and conceptualizing how networks were composed of different technologies: "While the system of computer resources is comprised of millions of hard drives, servers, routers, fiber-optic cables, and networks, we call it 'the cloud': a single, virtual object." This conceptualization is not just a simple metaphor but instead an example of how computer science virtualizes: "a technique for turning real things into logical objects, whether a physical network turned into a cloud-shaped icon, or a warehouse full of data storage servers turned into a 'cloud drive.'" Just as the "formation of water vapor, water crystal, and aerosol" in the sky is referred to as a cloud "to give a constantly shifting thing a simpler and more abstract form," the virtualization of cloud computing makes the heterogeneous technologies and networks abstract.[22]

The cloud appears in the interface as a "layered abstraction" that allows the user to store files without worrying about the physical material or location of the drive:

> The idea thus offers a spatial model of understanding and even standardizing computing: each layer depends on the more material layers "below" it to work, but does not need to know the exact implementation of those layers. Cloud computing is the epitome of this abstraction, a way of turning millions of computers and networks into a single, extremely abstract idea: "the cloud."[23]

Hu, in other words, explains how the cloud signifies the layered nature of the networked computer, or how it is an interface. As both a sign and signal, the interface is not just located between humans and computers but also between the different layers in the software and hardware, such as between different levels of programming, or different hardware in the computer or on networks.[24] Each interface level is an abstraction in the *mise en abîme* architecture of the computer, where interfaces are layered within other interfaces.[25] Behind each level of the interface hides another level. The application programming interface is not "truer" than the application, nor is the programming language Perl truer than the Linux operating system. They are simply operational instruments that allow the user to combine instrument and media. From the graphical user interface, to text-based command lines, to object-oriented programming languages, and so forth, interfaces allow for a combination of signs and signals at different registers of abstraction that in one way or the other, are combined with one another.[26]

Whereas the interface layers of the PC function within the confines of a relatively local system, though, the cloud interface along with its layers of computers, networks, and data is a meta-abstraction: it produces a general abstraction and absence on a global scale; all other interfaces are turned into a single and abstract idea of "a cloud." In this sense, the cloud interface is an interface that has become general (in everything) and global (everywhere). This globalization makes the interface pervasive, in the sense that an increasing part of the environment becomes part of and attached to the system, and ubiquitous in that it becomes increasingly difficult to determine what is not part of the interface.[27] The cloud—its smoothness and obstruction of intervention—has become an interface to the "clouded" multiplicity of global hard drives, servers, routers, cables, and networks. This virtualization of the architecture—its general as well as global abstraction and displacement—becomes a way to understand the architecture: by calling it a cloud, it allows us to manage the multiplicity of a complex architecture. It is an abstract metainterface.

Compared to the vernacular infrastructures of *Summer*, the architecture of routers, servers, and more is hence also displaced in significantly different ways. Where the architecture of the vernacular web becomes an organic entity of interfaces that reflect the practices and aspirations of the users who build it, the cloud architecture is of a different kind. It is still an architecture that people live in, but it is experienced as abstract, vaporous, and smooth. It is present as a smooth and vaporous reality, but is absent as a physical reality—out of human intervention. It is architecture for people to live in, but not for users to build.

The Cloud's Legacies: *Tantalum Memorial*

The clouding and alienation of the network architecture that come with the metainterface does not appear out of nowhere. As seen in the *Internet Machine*, the architecture is still there, somewhere, maintained and cleaned by someone. Paradoxically, the virtual architecture, the seamless network that encompasses the entire globe, builds on the values of the early Internet—such as the sharing of information—but the centralization and displacement of the architecture overshadows the users' ideals. The cloud has become a commercial platform for the transaction of data in a globalized world. As such, it subsumes and delivers the dreams of Internet pioneers, but also follows historical power strategies. As explained by Hu, the infrastructures of the network often originate in a national military apparatus, such as the US missile early-warning systems, command bunkers, weapon storage centers, and other phenomena reminiscent of the Cold War. The installation of fiber-optic cables also follows the railroads and has even deeper historical roots.[28] By looking at the longer history, and how the cloud relates to older national and military structures, it becomes apparent that the cloud—the fiber cables, data centers, and everything that globalization more or less is built on—is not a cancellation of spatial or material properties. Rather, it is a paradoxical reemergence of a sovereign power grafted on top of older and established infrastructures.

The London–based artist researcher group YoHa (Matsuko Yokokoji and Graham Harwood) has through various collaborations made a long list of interesting projects that explore how new technologies and networks are grafted onto older technologies, and the couple continually looks at the concealments of labor, class, (post)colonialism, and race. The sculptural installation *Tantalum Memorial* (figure 4.5) from 2008 (made in collaboration with Richard Wright) is, as Yokokoji and Harwood describe

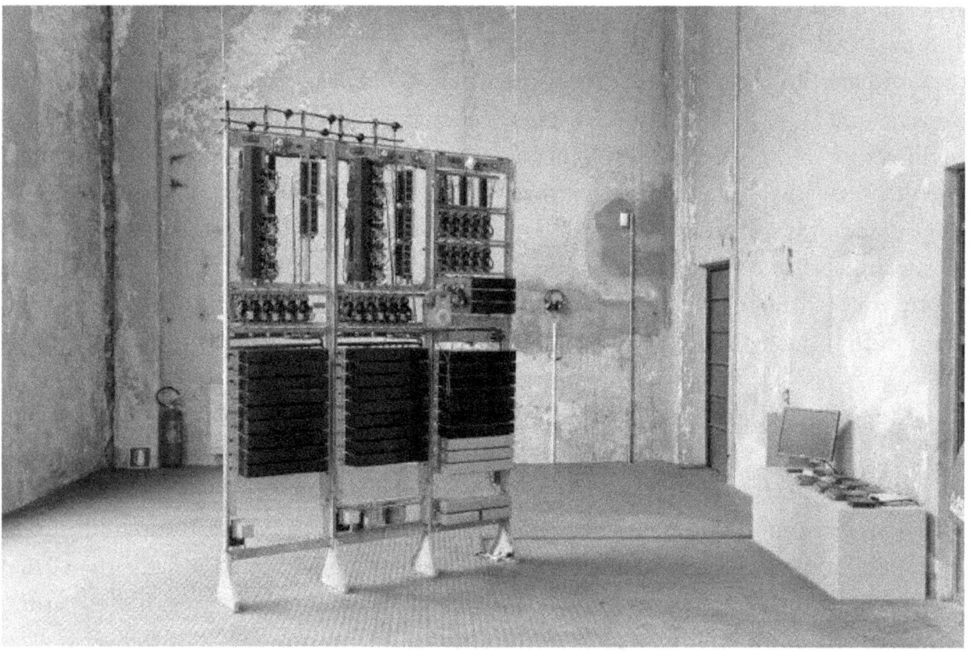

Figure 4.5

Tantalum Memorial (2008) by YoHa and Richard Wright. The sculptural installation is an electromagnetic telephone exchange connected to a temporary telephony network. The system builds on the Congolese tradition of "pavement telephone," a social telephone system. It calls Congolese listeners, plays them a phone message, and invites them to record a comment. Spectators can see the telephony switch and listen to recordings. The work is a memorial to the more than 4 million victims of the Congolese Coltan Wars. Courtesy of YoHa and Richard Wright. Photo by Matsuko Yokokoji.

it, a telephony-based memorial to the more than four million victims of the Congolese Coltan Wars.[29] Tantalum, a metal, is extracted from coltan ore. It is essential in cell phones, and is considered by many as one of the main origins for local and violent territorial conflicts.[30] The installation consists of an old electromagnetic Strowger telephony switch triggered by a computer that tracks calls from what is dubbed a "Telephone Trottoir" network, made for the Congolese diaspora in London. It is a telephone version of the Congolese "radio trottoir," or "pavement radio," which is a way of passing around news and gossip on the street in urban Africa to avoid state censorship. Telephone Trottoir, however, functions as a social telephone system that calls Congolese listeners, plays them a phone message, and invites them to record a comment and pass it on to a friend by entering their telephone number. In the installation, the spectators can see the telephony switch and listen to these recordings through headphones. *Tantalum Memorial* consequently combines a critical exploration of the mining of minerals in Congo, the culture and lives of Congolese immigrants who have fled from the conflict, and the social network technologies of the telephone as an alternative to centralized, commercial social networks. The installation hence makes visible the telephone technologies, their history, and the oppression that is related to the current extraction and production of minerals for their production, and simultaneously redesigns and uses the technology to form a new network: it is both criticism and execution.

Economic Fictions: *OCTO*

Another contemporary artist group that examines the prehistories and legacies of technical infrastructures is the Berlin-based art collective Telekommunisten. In its series of "miscommunication technologies," which includes *Deadswap, R15N, thimbl, Numbers Station,* and *OCTO,* Telekommunisten, like YoHa, focus on how these hidden material infrastructures are wound up in political economies, but they also point to the conditions of executing alternative networks. In general, the projects can be seen as a criticism of corporate network technologies and rhetoric, but simultaneously they are critical interventions into alternative ideas on free, open, noncorporate, and private communication networks. Often the projects do not just work as intended, even though the basic technology is in place and the dysfunctionality is part of the critical reflection. An example of this is *thimbl*, a decentralized microblogging service that works with the existing user information protocol called Finger.[31] Telekommunisten argues that *thimbl* is technically functional, but is an "economic fiction," too, which in order to succeed at large, needs financing on a similar scale as corporate social media.

Since there is no possibility to control its users and their data, there is no way to gather such funding. As an economic fiction, *thimbl* points to the economic conditions of the cloud: centralization and black boxing is promoted in order to capitalize on the Internet. Decentralized alternatives might well be technically possible, and in fact some are already functional, but they are not realizable under current capitalistic conditions in comparison to marketed, global cloud platforms. *Thimbl* illustrates that the interface industry of the cloud does not exist for purely technical reasons, but also—and perhaps mainly—because it is a competitive technological business model.

The largest of Telekommunisten's miscommunication technologies projects is *OCTO* (figure 4.6), which premiered in 2013 at the transmediale festival in Berlin and has later been exhibited in several places.[32] With two vacuum cleaners, one kilometer of plastic tube, a central station, and eight destination stations, the "Intertubular Pneumatic

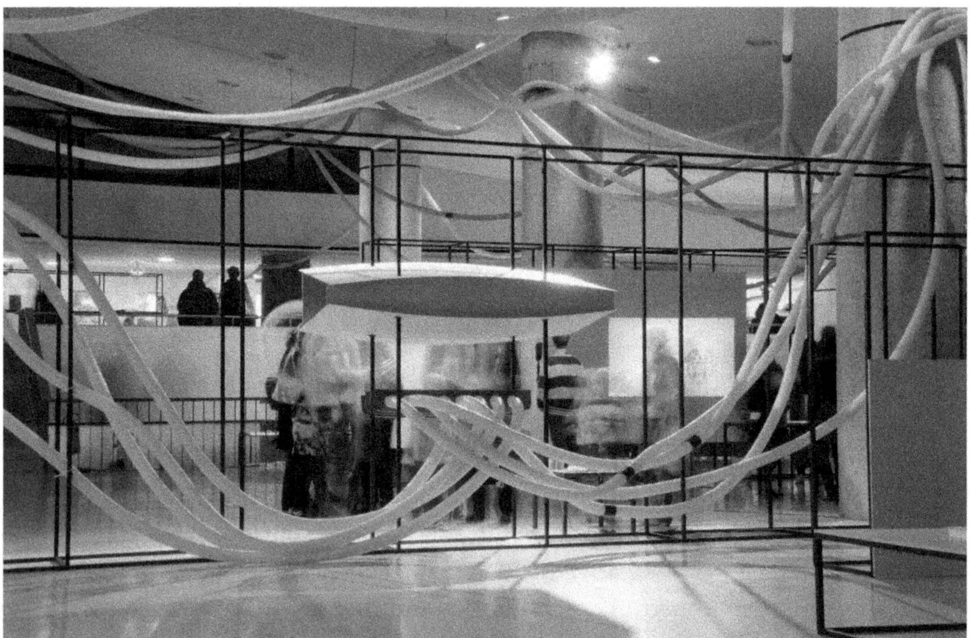

Figure 4.6
OCTO (2013) by Telekommunisten. An imitation of an obsolete technology, the pneumatic tube system ("Rohrpost"), used to transport capsules with solid objects in the nineteenth and twentieth century. Made by two vacuum cleaners, one kilometer of plastic tube, a central station, and eight destination stations. Courtesy of Baruch Gottlieb, Jonas Frankki, Dmytri Kleiner and Jeff Mann. Photo by Juan Quinones, transmediale (Attribution-ShareAlike 2.0).

Packet Distribution System" imitates an obsolete technology.[33] Pneumatic tubes that use vacuum for transporting capsules with solid objects or messages were introduced in the nineteenth century. For instance, in the local context of Berlin, the "Rohrpost" extended from 1865 to 1939 into a system of four hundred kilometers of tubes underneath the city's roads, and was still used until 1976 in East Berlin.[34] Pneumatic tube systems are, in fact, still in use in hospitals or other places where there is a need to transport physical objects, but first fax machines and later e-mail have (naturally) conquered most places.

Hence, Telekommunisten revives an old technology, as not only a reminder of a forgotten prehistory, and how new networks relate to former ones, but also an economic fiction. With this fiction (or "venture communism," as Telekommunisten chooses to call it), it points to how the materiality and politics of the old technologies are overlooked in the public promoting of the new networks. As Baruch Gottlieb and Dmitry Kleiner from Telekommunisten explain, "The utopian OCTO rhetoric is exuberantly cliché, promising all manner of human empowerment and positive transformation—and conveniently leaving in the shadow of bold promises the fact that this technology will be completely centralized and transfused with invasive security and monitoring technologies."[35] This centralization is demonstrated in both the corporate rhetoric of the venture start-up and technical infrastructure of *OCTO*, where all messages are handled by staff and have to pass through the central station.

Gottlieb and Kleiner describe this physical labor as a piece of "labor theater": "Unlike the Internet, where the physical labor is hidden, the labor in OCTO P7C-1 is presented as a central theatrical aspect of the work. The OCTO company provided the second layer, the social fiction, constantly driving home the lesson that there is a price for the convenience of every new technological utopia under capitalism."[36]

Through this labor theater, Telekommunisten stages how cloud interfaces hide labor, centralized infrastructure, and tracking behind utopian rhetoric. The rhetoric also hides the fact that the infrastructure is not just new but grafted onto older technologies, too—in this case, vacuum cleaners and plastic tubes. *OCTO* thereby questions the intentions within the construction of a phantasmagoric interface, it asks how certain structures of power and capitalism is reproduced and why the critical alternatives might not work. It does this by comparing the rhetorical construction of newness with a prehistory of old, and contrasting these historical technologies with the cliché rhetoric of newness. In this way, *OCTO* makes the audience reflect on how to see the new through the old, including the ways that new technologies are grafted onto old structures, ideas, and forms of capitalism, and how this is hidden by the clichéd rhetoric of newness.

The Cloud as the Phantasmagoria of Globalization

As Telekommunisten and YoHa's work and Tung-Hui Hu's prehistory especially point out, the cloud does not produce a deterritorial planet but instead continues old strategies of territorialization. In this sense, the cloud is not only an expression of virtualization; it is also a phantasmagoria: the cloud is a powerful shadow play (or a labor theater, to use Telekommunisten's words). Historically, phantasmagoria is an eighteenth-century illusionary and magic theater where images of demon and ghosts where projected onto walls or smoke, with the use of hidden magic lanterns. Yet the notion was used by Walter Benjamin, too, who extended Karl Marx's "statement on the phantasmagorical powers of the commodity to cover the entire domain of Parisian cultural products" in his exposé on the arcades project.[37] The commodity culture constitutes a phantasmagoria that embodies the deepest desires of its worshippers, but constantly hides its origins, and in this sense, the cloud, and the cloud's metainterface, can also be seen as a disguise of capitalism that operates on the level of social and collective dreams.

If globalization "takes place only in capital and data," and "everything else is damage control," as observed by Gaytri Spivak, then the cloud is the phantasmagoria of globalization.[38] In Hu's words, the control of time and space "indexes a *reemergence* of sovereign power within the realm of data."[39] With the cloud and its globalized networks, the world is turned into a planetary object with an air of infinity. Worlds have their ends, and are followed by new worlds, but planets are part of a different space and time specter. Globalization thus presents us with an ambiguity: our communication infrastructures are presented as virtual, global, and infinite, but as noted by both Hu (as a historian) and Lialina, Arnall, YoHa, and Telekommunisten (as artists), they hold in them also a world. Behind the phantasmagoria of the cloud, there are social, political, material, environmental, and territorial conflicts.

In other words, the virtual space (the globe) and virtual time (the smooth real-time streams) are something that is produced, and this production is seen in the tendency of artworks like *Summer, Internet Machine, Tantalum Memorial*, or *OCTO*. Each in their own way suggests how globalization has its world as well, and how the smoothness of streams and virtualization of complex networks are produced as tools of governance and economic or political power. Where network architectures used to be vernacular and social, they have now become out of human reach—presented to the user as a virtual infrastructure in a smooth and "not-to-worry-about" metainterface that displaces time and space: media are turned into streams with no apparent origin, and networks into virtual entities out of human reach. It is architecture for people to live in, but not for users to make, change, or in other ways interfere with. For the user of the cloud, this may be experienced as the freedom of mobility, "here and now," and unrestricted choice, yet it is no different than the phantasmagoric powers of former commodity culture. It is the shadow play of a globalized economy that seeks to hide its national, material, and capital interests. It is a technology that has oppressed its legacies as well as succeeded in presenting itself as a new reality that hides not only its infrastructures but also its political and real implications behind a veil of clichés and utopian rhetoric.

The cloud interface is a significant example of how globalization enters our desktops, screens, pockets, and environments—how globalization is inscribed into our lives and culture. These developments have an aesthetic where the cloud becomes an abstraction and extra layer in the interface, which is mainly experienced as a virtualization that displaces a material infrastructure. Consider, for instance, the function of a standard interface element, the button, and how it has gradually changed from a simple mechanical actuator on old machines to something like the "like," "search," or "buy" buttons in cloud interfaces. From being a mechanical connection in a machine, the button, with the graphical user interface, changed to become a metaphor for the execution of scripts: a sign that connects to signals as well as a medium that is an instrument. In the cloud interface, it has become an instrument to not only execute a script but also the production of data that can be read as signs of user behavior and exchanged in a network of interfaces: the button is a sign that not only connects to signals but to an endless stream of signals out there, in the cloud, too. The interface, as a metainterface, is not simply a way to address other local interface levels of software and hardware; it is a global network of servers, devices, and networks that are deeply integrated into apps, platforms, and web services. Consequently, the graphical user interface is no longer a mere user-friendly functionality that conceals the inner operations of the computer. Rather, it is a friendly face to a system of datafication tied to national and financial interests.[40] Through the aesthetic of contemporary metainterfaces, the interaction becomes virtualized, displaced, and *clouded.*

As globalization enters our pockets, the aesthetic of the cloud interface also forms the user's perspective on globalization. The metainterface as a cloud interface is globalization in our pockets, yet it is in our heads as well. The cloud interface is a way of sensing, understanding, and designing the contemporary global world. It connects the near and far in intricate ways, such as when users save their documents in a cloud drive like Dropbox, it is, as the company expresses it on its web page, "synced" to "data centers across the United States."[41] Near and far become mixed and convoluted in everyday tools and actions on the cloud interface. This phantasmagoria is rewarding and fulfilling for the user, who is able to share, participate, navigate, and more, in endless new ways, but it is also the origin of anxieties. Not only is interaction clouded;

so is perception. As Hu argues, a network is "always more than its digital or physical infrastructure. It is primarily the idea that 'everything is connected,' and, as such, is a product of a system of belief."[42] He even characterizes this belief as a state of desire, paranoia, or "network fever": "The cloudlike nature of the network has much less to do with its structural or technological properties than the way that we perceive and understand it; seen properly, the cloud resides within us."[43]

Therefore, the phantasmagoric process of clouding, or virtualizing and displacing a material infrastructure, is not only inscribed into the interface but also into perception—into how the world is read and understood as a metainterface. The remaining part of this chapter will discuss how this network fever, and the production of anxiety that follows from putting globalization into a pocket, function as a clouding of perception. The core assertion is that the metainterface not only affects how the global world is perceived but the way the problems and crises of this global world are configured, read, and understood as well. As an example, the chapter will extend the scope from the clouds of computing into the clouds of the sky, and deal with the most urgent global crisis, the climate crisis, but it will begin with a look at the endless war on terror. Simultaneously with the larger political dimensions—how the metainterface relates and configures the surrounding world—the metainterface will also be explored as a sensorium. Through analyses of artworks, the chapter will examine how the cloud interface configures perception from the sensorial machine of the war on terror to the ways humans perceive their environment under climate change. As a contemporary phantasmagoria and metainterface that encompasses the global, the cloud affects sense perception and experience of the world.

The Database as a Sensorium: *Endless War*

The cloud as a network-connected database is a system of belief, as Hu argued above.[44] Nowhere is this more evident than in the war on terror. In the attempt to target terrorists, endless amounts of data are constantly captured and analyzed by complex algorithms. One such data resource is the Combined Information Data Network Exchange (CIDNE) database, which is a collection of tactical information from troops, and developed by the US Military. The database became famous in 2010 when Chelsea Manning leaked documents from the CIDNE databases for Iraq and Afghanistan to WikiLeaks (the so-called Iraq War Logs and Afghan War Diaries).[45] In several ways, the database can be considered a "military cloud." The material contains reports from soldiers and intelligence officers, and specifies the time and geographic location of events that the US Army considers significant in combat, noncombat, propaganda, or other categories.

In addition, it includes fields that mark the number of casualties, whether they are allies, enemies, civilians, their reporting units, or more. As such, "The Afghan War Diary is the most significant archive about the reality of war to have ever been released during the course of a war," as WikiLeaks expresses it.[46] But what does such a massive archival activity reveal?

Many journalists have searched through this massive database (which contains more than seventy-five thousand documents) for important news or events in the material. The documents are classified secrets, and several of them exemplify lies, scandals, errors, and failures, such as civilian deaths, lack of success and increased numbers of Taliban attacks, foreign support for the Taliban, friendly fire casualties, and so on.[47] The Afghan War generally portrays a complicated and problematic war that has taken place in an unstable region, where it is often difficult to distinguish friends from foes.

The work *Endless War* by YoHa and Matthew Fuller (figure 4.7), however, takes a different approach to the CIDNE database. The work consists of three big projections

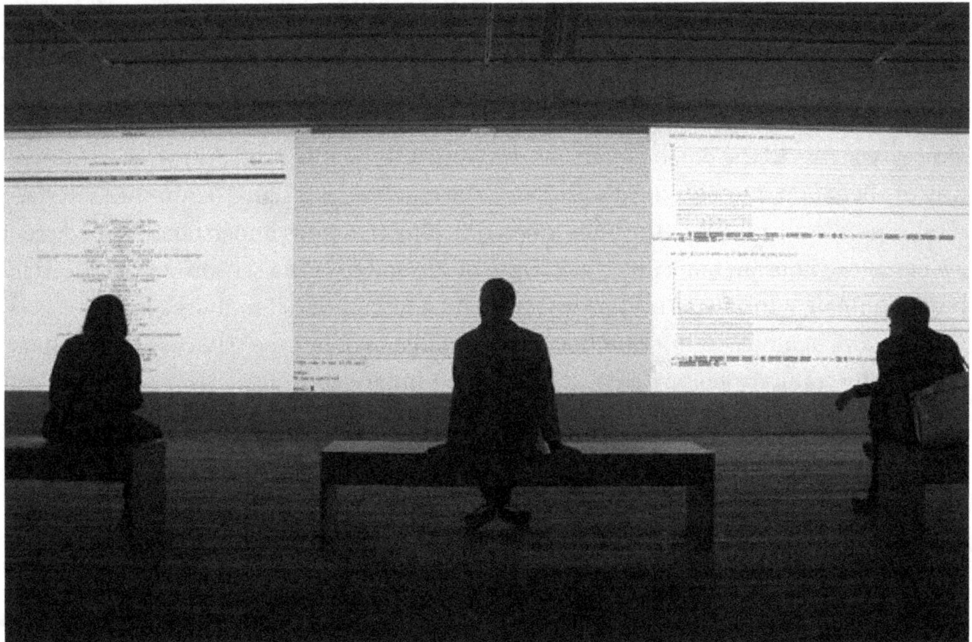

Figure 4.7
The Endless War by YoHa and Matthew Fuller. Three large projections of live searches and analyses of the more than 91,000 documents from NATO's war in Afghanistan (leaked by Chelsea Manning in WikiLeaks' so-called "Afghan War Diary"). The projections are accompanied by an audio track that sonifies the inner workings of the computer. Courtesy of YoHa. Photo by Paola Bernardelli.

of live searches through the database and an audio track that sonifies the inner workings of the computer. It is a noisy installation with flickering imagery of texts and lists that are generated by various scripts. For instance, the "N-gram fingerprint," which is a common algorithm that removes punctuation and spaces, changes all characters to their ACII representations in order to obtain so-called n-grams (a language model that predicts the next item in a chain).[48] The noisy sounds and endless strings of text symbols hardly make sense to any human being, and do not seem to reveal any journalistic sensations. Instead of providing a human–computer interface to the database, the installation attempts to display the interface inside the database machine (the signal–computer interface) that runs behind its military use.

This demonstration of an interior—how signs and language are read by machines, and converted into human noise—turns the gallery space into a chaos that not only metaphorically relates to the chaos of war but also represents the chaos of its informational architecture. In other words, it is a work that does not attempt to reveal the secrets of the CIDNE database but rather how war becomes a database, and how this alters the perception of war. Or in the words of the artists, it is a work that reflects the database as "a sensorium, an entire sensory and intellectual apparatus of the military body readied for battle, an apparatus through which the Afghan war is both thought and fought."[49] The flickering texts, and hum and noises of the computer, do not provide easy access to the database sensorium. In this sense it is a difficult work, which by way of alienation, presents a view into how the database as a signal–computer interface sees, orders, and understands. One does not obtain a new understanding of actual events in Afghanistan but instead a glimpse of the database as a sensorium produced by both a human, administrative bureaucracy and computational military cloud.

The work demonstrates a number of characteristics about the database as an "apparatus of the military body." First, the various military specifications, categories, and stamps of data express an ordered and bureaucratic way of seeing as well as representing reality. The CIDNE database provides representation schemes (casualties, civilians, allies, combat, propaganda, etc.) that are abstracted from the human perspective and instrumentalized by the machine. In this way, it regularizes expression through its grammar. As Matthew Fuller and Andy Goffey have observed in their book *Evil Media*, "Regularized expression points to the fact that for machines to address humans and things in *their* language, humans and things must address machines in *their* languages."[50] In other words, *Endless War* reflects such a formal human and nonhuman linguistic intelligence.

Second, the collecting and recording of so-called "significant acts," "psychological operations," and "counter-improvised explosive devices" are also run through a process of "data atomization," which "takes the intelligence report and deconstructs it, breaks its process down into a number of discrete parts by naming them as entities. The atomisation creates the possibility of both searching across domains and giving the queries scope. The other role for atomisation is the internal rendering of meaning to the system itself."[51] This atomization of data points to how the database and its endless capture of data becomes a sensory organ of the military system. The collection of data and their atomization (as in the n-gram fingerprint) suggests potential events that make way for new military strategies. The numerous reports are an idealized abstraction from the dirty messiness of "blood-letting and hardware maintenance and the ordering of toilet rolls," as YoHa and Fuller describe it. It is "a utopia of war" that "allows NATO's human souls to imagine themselves grasping the moment, the contingency of now. All of the war, all of the significant events, all of the time, all of the land, coming under the symbolic control of a central administration through the database, affording governance to coerce down the chain of command."[52] Put differently, *Endless War* shows how CIDNE transforms chaotic and faraway activities into an abstract military strategy (or at least is an actor in the transformation). This participation in an abstraction is not only technical (e.g., embedded in the military organization and its many institutions) but also an imaginary human feeling of contingency.

Third, the linguistic intelligence is executional. In a continuation of Philip E. Agre, the capture of data not only expresses a structured view of reality but the formal language for representing human activities structures activities, too; it is not only a grammar of organization and categorization but also a "grammar of action."[53] Following this, if *Endless War* reflects a formal human and nonhuman linguistic intelligence, it also shows how this has substantial ethic consequences. As YoHa and Fuller write, it is a military machine that explores "how the space between knowledge and power is transfigured to create a machine of death."[54] Perhaps no weapons are directly attached to the cloud of war (yet), but it functions as a way to survey and manage the activities of war, and its abstractions—which are cut off from the real chaos and technically instrumentalized in the military strategy—do have real-life consequences.

Database clouds and what they describe might not always be as sharply structured as in a military organization, yet in any cloud database there will always be a regularization of expressions: there will always be data that are atomized and detached from reality, but nevertheless make way for contingencies with real-life consequences. By using "The Afghan War Diary" as material, YoHa and Fuller depict a rather bleak perspective on the working and aesthetic of the cloud database, but they also make it clear how the database transforms behavioral inscription into strategic interfacing, or how representation and execution, human–computer interaction and signal–computer interaction, are intertwined. This strategic interfacing does not present a simple representation of the world; instead, it is rather integrated into a world where system workers (soldiers as well as smartphone users) partake in an augmentation of the world by technology (military intelligence technologies as well as Google's or Apple's services), whose execution is part of a reality (e.g., battle zones as well as urban life).

As an artwork, with the violence of the war looming in the title, *Endless War* reproduces cloud computing's tragedy of alienation. As the title indicates, it is an endless tragedy—parallel to the endless execution of the computer's scripted algorithms. It suggests that the cloud database is integrated with warfare, and it transforms the war on terror into a bureaucratic overload of information handling, which is difficult—if not impossible—to overview and understand. While this is a clear reflection of the military database, it should after all not be too hastily transferred to the understanding of other cloud databases, though the basic conclusion still holds: in order to understand how the executional database might work, it is necessary to realize how the world itself has become a data-producing interface.

Viewing *Endless War*, one might even ask if all these endless data make sense? Are there coherent plans and understandings for the war, or does *Endless War* ultimately demonstrate the workings of a nonsense machine producing a utopia of war that derailed politicians, public opinion, the military, and ultimately the war itself? *Endless War* in this sense points to the impotence of a utopia of military cloud computing when faced with complicated and highly contextual real-world conflicts. It questions whether the promised "smart" solutions of datafication fall short when the actions are more serious than choosing music or being offered advertisement.

In conclusion, *Endless War* illustrates how the world of globalization is perceived as a metainterface. This is not to be understood in its banal sense—that the world is interactive—but rather in the sense that the world is perceived as an abstraction (a "globe") that at once reveals and seals off a reality. The global world is one that produces data that can be fed to machines and read as signs of sealed-off processes. When the world becomes a metainterface, it produces a particular clouding of perception in the sense that perception of reality becomes saturated with contingencies: neither true nor false, but a proposition. Warfare, and the war on terror, is one aspect of globalization where this becomes evident; another is the climate crisis. If *Endless War* deals with the database as a sensorium and the inhuman database perspective on the war on terror, particular attention must be given to how the reality of potential crises and threats are perceived on a human scale, too.

The Cloud and the Climate: *Nuage Vert*

What does it imply to live in the phantasmagoria of clouds, and how does it look outside a strategic database perspective? As discussed in the introduction, in relation to Joana Moll's *CO2GLE*, cloud interfaces are not only global but also have planetary consequences.[55] A Google search appears in a fraction of a second, and the whole world of information is at the tips of the user's fingers; nevertheless, this has physical consequences. Arguably, compared to the many stand-alone platforms distributed across the planet, cloud computing and the centralization of computer storage potentially may reduce the overall emission of carbon dioxide from computers: shared data centers can employ servers to get an equivalent capacity of storage; because of the large scale, cloud providers can take advantage of more efficient ways of controlling temperature and humidity; and so forth. Cloud computing is therefore often advocated as a green technology.[56] The arguments may seem convincing, although the estimation of the overall effect of cloud computing on the environment is naturally a complex issue that also needs to question whether it leads to an increase in demand for computing, alternative behaviors, and much more. On a perceptual level, however, which is another level of *CO2GLE*, cloud computing implies that consumption and production is displaced as well as clouded: it is somewhere else than at the tips of our fingers.

This displacement of pollution seems to be an overall trend. Consider, for instance, the exhaust from cars, which has led to smog problems in every large city for decades, and how it is now filtered and made invisible. Particle filters for diesel cars, for example, definitely have a clear and useful purpose, but exhaust is still harmful. As the Volkswagen scandal in 2015 demonstrated, though, the harmfulness of emissions is no longer directly perceptible. In fact, it is something that is mainly perceived by artificial sensors and computers that run complex software. To subvert the emissions test, Volkswagen programmed software that would activate emissions control under circumstances that resemble the laboratory testing protocols. In other words, during testing (or similar circumstances), emissions were regulated to meet the required standards. To reduce the emissions, Volkswagen merely had to reprogram the software of its cars. Or to put it another way, the real emissions were no longer directly sensed (by human or artificial sensors), but something that was validated in the code.[57] Reality has become mediated in software, and the control of reality inscribed in code, though of course the real emissions from cars are still deadly harmful.

In addition to Moll, the Paris-based environmental artist group HeHe (Helen Evans and Heiko Hansen) has explored how pollution is removed from the visible and sensible (including its playful experiments with car emissions long before the Volkswagen

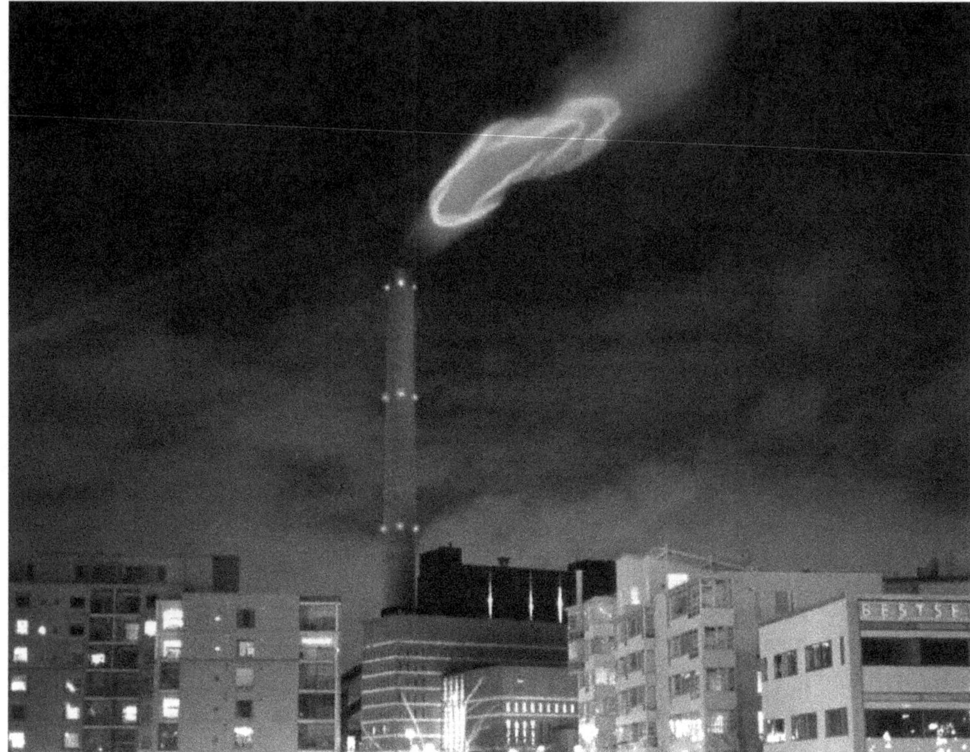

Figure 4.8
Nuage Vert 01 (2008) by HeHe. A green laser-drawn cloud onto the chimney smoke of a Helsinki power plant. The size of cloud relates to the production/consumption of energy: the lesser the consumption, the larger the cloud. Courtesy of HeHe. Photo by HeHe.

scandal in *Toy Emissions* [2007]). In its work *Nuage Vert 01* (2008) (figure 4.8), HeHe projected a green laser-drawn cloud onto the almost-invisible chimney smoke of a Helsinki power plant.[58] The size of the illuminated cloud was correlated to the power consumption in the city and power production at the power plant. Nevertheless, and quite symbolically, the relation between the size of the cloud and consumption was reversed, so less consumption resulted in a larger green cloud. Subsequently, the citizens of Helsinki could see their power consumption directly: the less they consumed, the more green and beautiful a sky they would get.

In this way, HeHe's work is not a simple interface for saving energy; it is not just an interface that makes the citizens see what is hidden. As an interface to emissions, *Nuage Vert 01* resembles the smart interfaces of quantification that through visualization and rewards, encourage their users to behave better. In fact, the installation resulted in a small reduction of emissions during the time it was set up. Yet the interface also explores the crisis of perception related to the cloud interface: not only does it make us see what we cannot see, but also *that* we cannot see. As HeHe phrases it, "We are the Google Earth generation. We need to observe everything from above, from afar, through the glasses of a scientific model."[59] In this observation "from above," the green cloud is a collective and reversed cybernetic game that is technical and "smart," but the work's significance cannot be estimated in megawatts alone. Its real importance lies in how it makes the citizens see the relations between energy consumption and carbon emissions—as something that is imperceptible, unless there is a technical mediation.

Climate Interfaces: *Toxi•City*

The vision of the planet from above and afar is a realization of a collective possibility for social and political action. Viewed from above and afar, the climate—as the constant interactions of humans and nonhuman actors in a system—may potentially become present and vernacular. Living in the climate crisis, however, the call for uncompelled action is not self-evident. "What can you do?" and "how can you see it?" seem to be the proliferated questions.

The US artists and researchers Roderick Coover and Scott Rettberg (the latter based in Norway) have explored this aspect of the climate crisis in their recombinatory film *Toxi•City: A Climate Change Narrative* (figure 4.9).[60] The film portrays people living in a river estuary after the climate crisis has struck with repeated storms and floods. It takes place in a near future in the postindustrial Delaware River Estuary, which is home to several large oil refineries and a nuclear power plant, and located near large cities like Philadelphia and the coastal shores of New Jersey. The work is usually displayed as an installation in a dark room, on a large 5:1 wide-screen cinemascope. The images are of the Delaware River along with its waters, ships, shores, and industrial architecture. The wide panoramic format, with rarely any humans in the picture, emphasizes the power of a nature out of balance. This impression is sustained by low-key colors, semitransparent impositions of ships and water, and slow camera pans that emulate the river's floating. The imagery has its own beauty in spite of the postindustrial decay, equaling what in the German Ruhr district has been called "industrial nature."

Toxi•City's narrative combines fictional testimonies about how to survive in this postcatastrophic environment with accounts of actual deaths caused by potential storms and floods related to climate change, such as Hurricane Sandy in 2012 and the still more frequent and devastating hurricanes such as in the unusually hyperactive

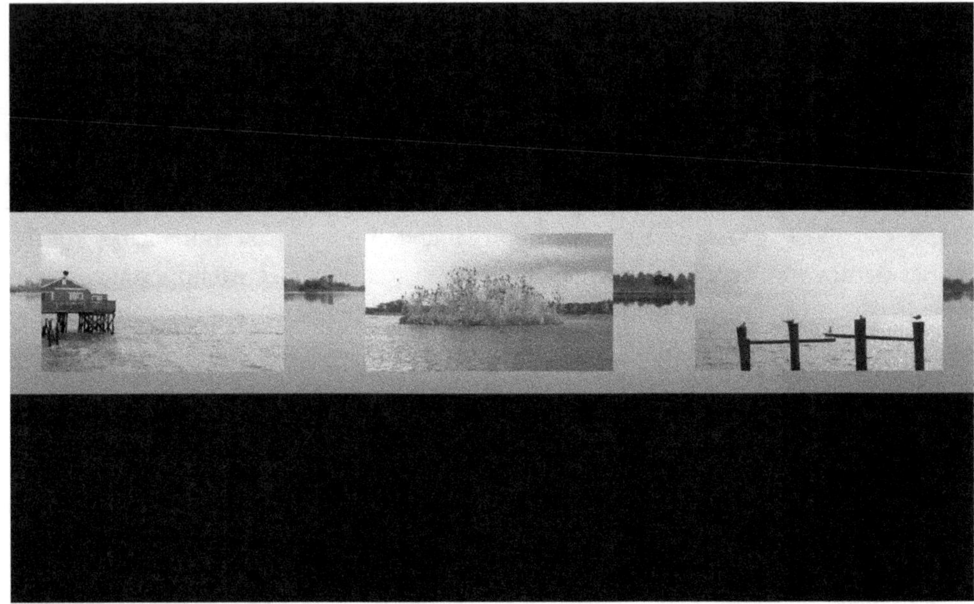

Figure 4.9
Toxi•City—A climate-change narrative (2014) by Scott Rettberg and Roderick Coover. Recombinatory panoramic fiction film that portrays the industrial nature of the Delaware River Estuary, and the people living there. The movie is a post-human narrative of how humans struggle to both interpret and survive a climate crisis, but also a movie of how to narrate this crisis. Courtesy of Scott Rettberg and Roderick Coover. Still from movie.

2017 Atlantic hurricane season. The film has four beginnings, thirty-eight episodic narratives from its six protagonists, thirty-three nonfiction death stories from actual storms, five segues showing the urban and river landscape, and the ending.[61] The episodic narratives from the six primary characters play in semirandom order, only intervened by the factual death stories and segues.[62] The beginnings anchor the stories in the postcatastrophic estuary landscape, and its many episodes introduce the daily struggles of the six main protagonists at a place where disastrous recurring floods and storms—and the pollution, shipwrecks, and faulty rescue operations that follow from them—have become a condition for living.

The mixture of nonfiction death stories, the actual setting, and fictive near-future stories on how to survive and adapt creates a topographic narrative of a drowned, polluted, and diseased landscape where people struggle to get on as well as understand what they are exposed to. For instance, a pig farmer suggestively asks if all the warnings from "tree-huggers and democrats and Hare Krishna and so-called scientists" made "us one bit more prepared?" He seems to be the only character who somewhat succeeds in this tough environment; as he says in typical free market thinking, "Show me a crisis and I'll show you an opportunity."[63] The other protagonists are mostly victims of their helplessness and are seeking signs in nature.[64] Even the beautiful sunsets are a cause of anxiety for the young woman when she realizes that their many different hues, which she enjoys so much, are probably due to pollution. Still, she finds them "kind of magical" and settles into resignation: "Life is hard but I think it is sort of our job to live in the moments that we have now."[65] Though the protagonists strive to understand the situation, their potential possibilities for taking action suffer. It is a desolate story about the loss of lives and nature, and the intervening death stories are like the chorus in a Greek tragedy. Nonetheless, there are also scattered moments of connection and humanity, including the final episode where the young woman sees the teenage boy digging a grave for his mother: "I walk towards him. I know what he is doing. I can help him with this task."[66]

The feeling of human powerlessness is emphasized by the poetics of the recombinatory database narrative. Instead of being sequential, the narrative is a paradigmatic landscape of voices without much order, besides the fact that they live in the estuary and are victims of its ecological breakdown. The narrative cause and effect usually delivered by an explanatory human narrator is replaced in *Toxi•City* with an algorithmic narrative mechanism, where there seems to be no order, understanding, or interpretation apart from the ordering of beginnings, character narratives, death stories, and segues. The random conditions of the machine and its unintelligent, combinatory algorithm reflect the desolate fate of the protagonists in an ecology whose balance has been profoundly disturbed by humans, to the extent that they literally drown in their own toxic remains: it is a posthuman narrative told by a posthuman machine.

The posthuman narrative of how we struggle to interpret, act on, and even survive within a climate crisis is also a story of how to narrate this crisis. The characters are generally not able to get a general perspective but instead are enclosed by the crisis environment, and adapt their modest strategies trying to survive and make sense. *Toxi•City*'s own algorithmic poetics and panoramic, cinematographic aesthetics, however, work as a way of telling and sensing the climate crisis. This creates a tension between the helpless individual destinies of the estuary, on the one hand, and aesthetic and poetic qualities of the artwork, on the other. Perhaps it could be characterized as an anthropocentric panorama, a blues on the climate crisis that depicts the catastrophes and sorrow, but also allows the viewer to indulge in it, to feel the pain and basically to be engaged, at least aesthetically, and in this way, gradually explore what a panorama of climate change could look like.

Panorama literally means "all seeing," and as the first visual mass medium it became popular in the nineteenth century as a phantasmagoric architectural medium in which the mastery of the modern bourgeois perspective of control and commercialization is both learned and glorified.[67] In Coover and Rettberg's work, it becomes a way of experiencing what the characters in *Toxi•City* cannot quite see. It is a climate change panorama that demonstrates how nature and the environment are controlling the characters, by being out of control. As Coover writes elsewhere on the panorama, "The viewer has the impression of seeing everything—of being able to grasp the image as a whole. Paradoxically, the inverse may be more accurate. The panorama offers the viewers an illusion of commanding a total view of a moment; actually, it is the image that encompasses the viewer in the exotic locales of its form."[68]

In this way, *Toxi•City* can be understood as a continuation of Benjamin's efforts to, as the acclaimed US Benjamin scholar Miriam Hansen expresses it, "reconceptualize experience through the very conditions of its *im*possibility."[69] As a panorama, *Toxi•City* demonstrates the limits of human perception: the circumstance, as also experienced by the narrators, that climate change cannot be seen clearly. But it does so as an all-seeing technological, narrative spectacle: a panorama that depicts the human imprints on a geologic scale. In this panorama, the spectator loses control of the image. It is a panoramic illusion of total overview where the image encompasses the spectator. Hence, if *Nuage Vert 01* shows that we cannot see the climate, *Toxi•City* illustrates both *that* we cannot see and *what* we cannot see: the climate is clouded, displaced, both everywhere and somewhere else. The possible spaces for seeing and taking action are limited, but *Toxi•City*'s panorama, the technical mediation of climate change, presents itself as an opening—or rather, as a way to reflect a "clouded perception" of reality.

Weather Interfaces: *Snow*

As *Toxi•City* exemplifies, the human imprint on the planet alters perception: it makes perception work on a different scale. Imprints of consumption and production may be seen in smog and other kinds of pollution on a local scale, but are often not perceivable. They work on a different scale, where causes and effects are not seen in the everyday but instead on the level of the planet's geologic eras. Seeing the environmental conditions in the everyday is hence simultaneously an experience of losing sight. A call for action may be directed toward the environmental activist, but it is equally important to reflect how reality is produced on the level of perception. If *Toxi•City* speculates on how one may live with this change of perception in a time when climate changes are beginning to have direct impacts on the everyday, the US poet Shelley Jackson speaks

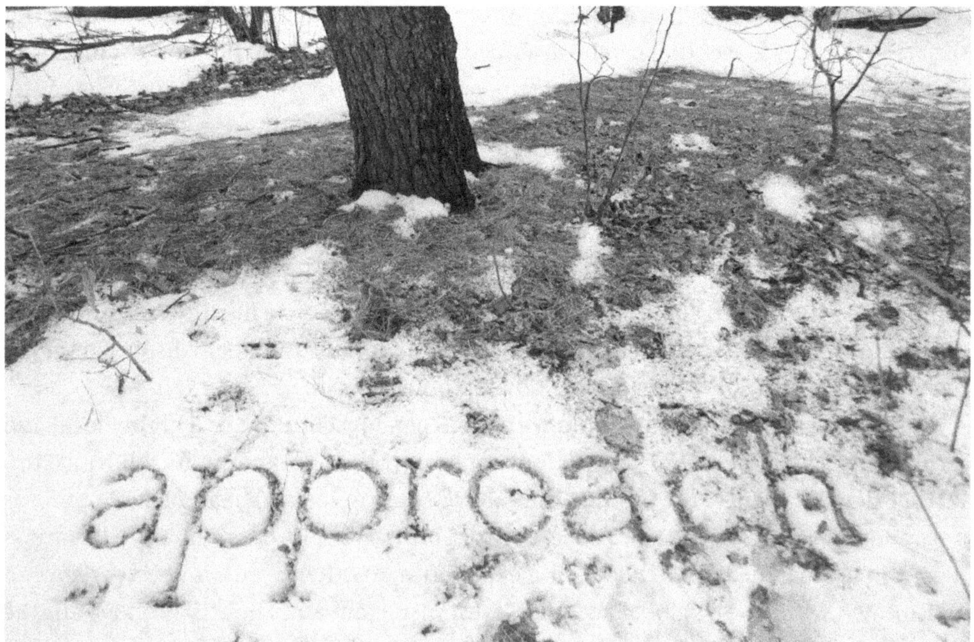

Figure 4.10
Snow—A Story in Progress, Weather Permitting (2014–) by Shelley Jackson (photos from Instagram). Story published over years as images of words written in snow, distributed on Instagram and Flickr. Snow becomes a writing surface with its own material poetics, but the presence/lack of snow for writing also points to how the weather becomes an interface for climate change: a potential sign of an otherwise imperceptible phenomenon, and how this becomes an emerging concern, spreading across the social network. Courtesy of Shelly Jackson. Photo by Shelley Jackson.

to the perceived nature of imprints along with the inscriptions themselves in a time of metainterfaces and cloud computing.

Jackson began publishing her story *Snow* in January 2014 (and the work is still under publication at the time of this writing). It is published sequentially as images of words written in snow (figure 4.10) that are distributed on Instagram and Flickr, and since every image only contains one single handwritten word, it both requires quite some work and snow on the ground in Jackson's hometown, Brooklyn, in order to be written and published.[70]

Snow's content is a phenomenology of snow in its many different varieties, how it surfaces the world, and even a record and exploration of the world: "There are depraved snows that make unwelcome advances and cerebral snows that sifting along surfaces seek knowledge of the countless forms of the world."[71] Snow becomes a writing surface, a kind of paper. Or rather, paper, the snow on the ground, and potentially many other surfaces appear as spaces that can be inscribed with signs. Jackson has, for instance, also published her work on human skin—tattooed on the skin of 2,095 volunteers as a living and mortal novella.[72] *Snow* expresses a fundamental relationship between writing and architecture, but naturally relates to a history of land art, too.[73] In contrast to writing in stone or even writing on paper, though, writing in snow is profoundly ephemeral. Or rather, it speeds up the ephemerality of inscriptions that is also present in tattoos (even real ones) and writings on paper. But like architecture, its solidity makes it immobile and related to local context. Reading the text, one literally reads not only about snow but also the different kinds of snow and locations that are displayed on the images (slush, frost, snow on streets, snow on objects, etc.).

In these ways, *Snow* shows a world with ephemeral inscriptions in a highly local and solid context. To reproduce *Snow* will inevitably abstract the text from its solid and contextual element. Hence, the proliferation of the work on Instagram and Flickr produces a paradox. Though the text is readable through images on Instagram and Flickr, it is meant to be imagined, too. The reading experience, in other words, is an experience of a dislocated writing. The global accessibility of Instagram and Flickr is at odds with the ephemeral quality and site specificity of the snow in Brooklyn, and reading the text on social media almost feels like a violation of the writing, like abstracting superficial data from the body of the world. Consequently, *Snow* explores and reflects on the underlying crises of representation and mediation in the frictions between the local, contextual, and ephemeral, on the one side, and the global and networked accessibility, on the other. It is, in short, not just a work in snow; it is also a work that displays the frictions between the site-specific text and the distributed photos and text on social media.

Yet the letters inscribed on the white surface of the snow bear more than a metaphoric relation to the printed letters on paper, and the crisis of representation that it expresses is much more than a mere dislocation of writing from paper to the clouds and metainterfaces of social media. *Snow* is a "story in progress, weather permitting," as is stated on the project's Instagram account.[74] In other words, it relates the material inscriptions of the story not only to social media but also to the weather. Snow, along with other weather phenomena, like wind, temperature, humidity, and clouds, are all signs of the climate. The changing shapes of clouds, increasing winds, decreasing temperatures, droughts, floods, and so forth, are thus signs of potential changes in the climate—changes that one easily loses sight off, as demonstrated in *Toxi•City*.

As *Snow* indicates, climate change has added a new semiotic layer to the experience of the weather. The weather is not only a conversation starter, or a forecast that can be used to take precautions for how to dress, when to seed, what routes to navigate on the ocean, and so on. The weather is extended from a local to a global scale of the earth and its geologic eras. The weather is, in other words, not something simply "out there" but instead has become an interface of contingency: not true, not false, but a potential sign of climate change. When the signs of climate change are observable as melting ice, floods, drafts, fires, and other changes in weather across the globe, global measurements and future models also become the foundation of discourse: the reality of the metainterface becomes a concern. For instance, in *Snow*, the lack of snow is a persistent underlying theme. The work unfolds over years, and as a reader, one constantly reflects on its presence as a sign of much larger weather systems. Is the presence of a snowstorm, or unusually warm weather and lack of snow in the winter, a sign of global warming?[75] Such questions are intrinsic to the reading of *Snow*, so diligently dependent on the weather.[76] In this way, *Snow* is a work about not only snow but also the epistemology of climate change—about how we perceive (the lack or excess of) snow in the urban environment as an emerging global concern that is spread across networks. The reader not only reads the meaning of the words; they read the writing of climate change in the local weather, on the surface of streets, or in the melting ice on the lake as well. *Snow* seems to suggest that the writing of climate change is as frail and insubstantial as melting snow on an urban pavement—something that slightly disrupts and slows traffic, but is soon cleared for efficiency. Climate change is staged as an important writing to read before the moment of reading is gone, and it is too late.

Cloud Interface Criticism

The cloud interface, as a metainterface, has been described through different axes in this chapter. It is experienced as an abstraction of an infrastructure of data and software. This abstraction has been incorporated by the platforms of an interface industry that is recentralizing the once decentralized structures and architectures of hypertext and networks—while simultaneously hiding the existence of such structures and architectures. Hypertext and the construction of vernacular networks was once seen as a liberating technology that allowed readers to choose between nonlinear narratives and even making their own links, but has now become instrumentalized by cloud platforms that seek to capitalize on the behaviors and language of their users (e.g., as demonstrated in Lialina's work *Summer*). Clicking and linking has become liking and sharing, and the users' interests are interpreted and processed into endless autogenerated recommendations and newsfeeds. The metainterface industry is increasingly interpolated as a metainterface between the user and their software and data. Whatever these data may be (music in a streaming service, data in a cloud drive, or biographical profiles on social media platforms), they are rarely accessible, controllable, or removable for the users, and generally the users have no way to precisely determine the location or full extent of the data in the cloud, or how they function and are interpreted.

As argued, the cloud is a sign of how the metainterface industry is a generalized and globalized phantasmagoria, a shadow play of smoothness where media seems to appear from nowhere. This is not only professed in the human–computer interface. The clouding of a material infrastructure is also a clouding of perception: it is a way of seeing and reading the world as an interface that is typical in a globalized world. Both the war on terror and climate change are examples of how the world is perceived as not only interactive (as a human–computer interface) but also clouded and contingent (as a signal–computer interface). The at once smooth and clouded interface is increasingly becoming a metainterface to reality that makes its users see their environment through data networks and algorithms, and their inherent models of potential behaviors of people or climate systems. The metainterface becomes a sensorium in and to the world.

The general analytic model in this chapter has been a focus on the process of virtualization and immaterial abstraction. On the level of the signal–computer interface, the objects of the world are turned into logical, computable objects that can be read and sensed by the machine (*Endless War* is a good illustration of this). This process of datafication, however, becomes part of the human–computer interface as a layered abstraction, too, where clicking an icon on the screen means interfacing with an endless number of computers, networks, and users globally. If the icon on the traditional PC typically points toward localizable software and data on a hard drive, it now often enacts a signifying chain of global networks and services that is commonly as well as metaphorically referred to as a cloud. In this way, the cloud interface has become more abstract than, for instance, the traditional interface of the PC, but it has also become more generalized. As a cloud in the sky, the cloud interface is more difficult to demarcate and delimit, and it has become more pervasive and ubiquitous. Yet such abstractions in the cloud interface are intrinsically dependent on a displacement of a material architecture. Rather than being the object of recursive negotiations between its users, the data centers are left out of public reach. Although the architecture does not appear in the interface, it is nevertheless a territorial materialization of sovereign power: its infrastructures are grafted onto existing infrastructures of railroads and military command centers (as described by Hu).

The perspective on the cloud interface can be narrowly (techno) logic, focused on solving clearly defined issues (such as climate change or terror). It may also be sociological and cultural with a focus on the larger coherences and impacts of the cloud on behavior, governance, ethics, and so forth. The cloud is a technology that is territorializing, but also holds potentials for collective actions (such as WikiLeaks) to understand global issues such as armed conflicts, and arguably data and models of climate change are more important now than ever. This chapter has sought to outline how artistic practice may explore the cloud. The presented analyses have been a way of looking at the rhetorical mechanisms of the cloud, and what is produced through its combination of abstraction and materialization. What hopefully has become evident is how artworks reveal their material tendencies, how they themselves employ and reflect a world of clouds and their politics in a globalized world, and how the world of clouds is not only a technological novelty but (like former media technologies) also involves a change of perception. The aim has been to critically discuss the politics of cloud computing, and yet how this relates to a sense perception that exists in the intersection between the human and nonhuman. The last chapter of the book will raise the question of how to employ such tendencies by design, to open up a new and more nuanced design of cultural computing.

5 Interface Criticism by Design

"You Say You Want a Revolution"

This book has described and explored the metainterface, a new interface paradigm that supplements and substitutes the interface of the networked PC. The intentions behind it have been to provide a critique of contemporary interface culture, and how it is a designed and produced reality. Interface criticism is not a new thing. Although the computer interface has a longer history, it gained particular cultural importance in the 1980s and 1990s. In Ridley Scott's famous advertisement video for the first Apple Macintosh computer (broadcast during the 1984 US Super Bowl finals), a blond, athletic woman runs into a hall where gray, expressionless zombie humans are marching and mindlessly watching a large screen. With clear references to George Orwell's *Nineteen Eighty-Four,* she is pursued by the thought police, but throws her hammer into the screen at the exact moment when Big Brother exclaims, "We will prevail!" A comforting male voice takes over and reads a message: "On January 24th, Apple Computer will introduce Macintosh. And you'll see why 1984 won't be like *1984*."[1] The Macintosh is the human voice that contrasts the manipulating dictatorial enemy. The 1984 Macintosh marks the commercial breakthrough of personal computing with a graphical user interface—a world that responds to the users' needs and desires, contrary to corporate computers and the command line operating system of IBM's PC that was its main contender on the market at the time.

Apple's *1984* has played a pivotal role in reiterating and adapting Orwell's dystopic warning about the context of the networked computer. The new graphical interface introduces a culture that seeks to liberate itself from an authoritarian surveillance culture as well as the prevailing understanding of the computer as a centralized control system operated by men in white lab coats in large corporations or the military. For instance, the US Army's Semi-Automatic Ground Environment (SAGE) control system—a network of local radar stations and large computers built by IBM, and operational from the 1950s up until the 1980s—was built to control North American airspace. The data from the radar stations were used to produce a map of a territory with live information about flight tracks and automatic calculations of possible interceptions of the fast-moving potential enemy jets.[2] In other words, Apple and other parts of the Californian computer industry and research community successfully reinterpreted the interface's monitoring element as a liberating revolution and hip tool for self-expression, rather than the end of the free world. The PC and graphical user interface alters the perception of monitoring, and paves the way for the interface as a new cultural form that users willingly interact with and adapt to in their own self-exposure.

Thinking of interface culture today, the user's interface has apparently been overly successful. As also argued within the field of human–computer interaction (HCI), interface design has shifted its focus from applying rigid guidelines and formal methods in the workplace to including culture, experience, and meaning making in our everyday life and culture.[3] Computing has become a cultural thing. Software and hardware companies produce the platforms of the most common cultural activities, such as listening to music, sharing images, and much more. Likewise, it has given rise to new cultural activities: new ways of writing, archiving, sharing, and so on. Even spatial practices in urban spaces and the perception of globalization seem to be influenced by the smooth functionality of an interface that is at once absent and abstract, but in everything and everywhere, and perhaps in the end not that different from SAGE's big data control system.[4] This book is built on the need to critically understand the new affordances of metainterfaces, and question not only what they afford but also the kinds of realities that they produce: a critique of the reality that Apple and others produce when computing becomes cultural and the interface a metainterface.

A common assumption of the user interface is that it produces emancipation. This is not only found in Apple's famous commercial and the self-understanding of a participatory "prosumer" culture but also in former visions of, for instance, hypertext, such as Ted Nelson's famous *Computer Lib / Dream Machines*, which already in the mid-1970s described how a shared writing space potentially may overcome the problems of linear text forms, and lead to liberation and new imaginaries. Such assumptions, however, may be seen in a larger historical perspective as well. According to the German media theorist Hartmut Winkler, the development of media is "deeply rooted in a repulsion against arbitrariness" and "long line of attempts to find a technical solution to the arbitrariness" that date back to the visual technical media of the nineteenth century.[5] The development of modern media technologies tends to oscillate between, first, seeing the new media technology as a postsymbolic ecstasy that is not only an ecstasy of the media but also of the ways of living it brings about, and second, a state of hegemony in which the new media becomes habitual, and third, a state of crisis where the arbitrary character of the medium's signs is rediscovered.[6]

A reoccurring theme throughout this book has been to uncover how dreams of technological and emancipatory futures are instrumentalized as well as subsumed under capital and other interests. As early as 1991, the hypertext author Stuart Moulthrop noted in his essay "You Say You Want a Revolution" that the responsibility for changes in the magnitude that Nelson and others dreamed of would lie in the hands of a diverse elite of software developers, academics, literary critics, legislators, and publishers, who in the end will remain faithful to the institutions they represent (i.e., the book, library, university, print industry, etc.). They will, for instance, have a clear interest in intellectual property, and therefore it would seem "equally possible that engagement with interactive media will follow the path of reaction, not revolution," Moulthrop writes.[7] When looking at the metainterface, and the services and platforms of not just Apple but also Facebook, Google, Amazon, and others today, it is obvious that the emancipatory dream of the interface is accompanied by strict control mechanisms that even follow the schemes of the former information architectures of, say, military computer control centers.[8] One may speculate if the time of vernacular webs, peer networks of PCs, and servers will one day appear as only a brief moment in the larger history of computing, where the centralized control of big data draws the larger line. Today, the cloud's client-server relations, apps, tablets, and other infrastructures of the new interface culture are driven by military as well as capital interests that turn the promises of the emancipatory interface—a world of free search, access, and sharing—into a phantasmagoria.

Following from Winkler's work, another ambition of the book has been to rediscover the arbitrariness of the interface. The metainterface has become habitual, and all aspects of life seemingly have become permeated with its grammar (from cultural production and consumption, to urban space, and even the perception of a global world). Even though the metainterface speaks a new and seemingly nonarbitrary truth about the world, the intention has been to demonstrate how arbitrariness finds its way, and how the metainterface functions as a language with a technological grammar. Hence, the rediscovery of the interface's arbitrariness in a time of the metainterface is far from driven by reactionary dreams back to a world of "innocent" media (back to the "old school" user interface and a vernacular web). Rather, the intention has been to show how the metainterface brings with it a language that can be performed in endless ways. If the net and software art of the 1990s and early 2000s illustrated how the human–computer interface was not just a window to new hyperrealities but instead a sophisticated language that combined human representation with computation (signs with signals), the artistic expressions of the metainterface today additionally relate to how signals are exchanged within computer systems. In other words, they reveal a tendency in a nonhuman language that builds on the massive collection and calculation of data—and how contemporary computing produces new cultural industries, new territories, and new realities. Not taking this reality for granted, but sensing it as a language and construct, may allow for other experiences of surrealism, beauty, anxiety, laughter, and much more (as the artworks brought forward in this book have hopefully demonstrated). This is a much-needed element of technology critique.

A central argument has been that to give way for such freedom, one needs to critically understand how the metainterface produces realities, or in this case, how it produces a new cultural metainterface industry, urban territories, and perceptions of a globalized world—including the perception of the endless war on terrorism and climate change. But how can this critical sensibility to the interface's tendency be used as a strategy for interface design? As this book has asserted, the design of metainterfaces often hides how the technical infrastructures are political and cultural constructs that give rise to substantial changes in both industry, the contestation of space, and even human perception. Interface design is, to put it another way, not just the design of smooth use and interaction but also the design of users and interactors along with the social and political reality around them: it is not just a design of a tool but rather a larger technocultural apparatus. In a similar way that critical HCI in the 1980s not only provided simple guidelines for better interaction but also actively sought to reflect the introduction of computers to the workplace, critical interface design today needs to reflect the realities produced by the metainterface. Given the metainterface's production of realities, it is important to take the next step and involve criticism by design, as an approach and element in interface design. Crises and conflicts as well as questions of access, ownership, profiling, and even changes in institutions and users' daily life are occurring with metainterfaces, and this should be reflected in the design of the metainterface. Succumbing to a smooth design would be manipulative in its ways of hiding the political and cultural dimensions along with how they are related to the technical.

This final chapter of the book—drawing on critical HCI as well as network and software cultures—will seek to outline how a critical perspective on the metainterface may exceed the analytic (as readings of a tendency within the artwork) and fuel new technological imaginaries, or in short, how it may function as a catalyst for a design paradigm.[9] *The Art of Platforms, Cities, and Clouds*, the subtitle of this book, points to how computing presently makes way for new artistic expressions, but also to how computing and interface design itself may be practiced.

Criticism by Design

Within interface design, there is a long history of criticism. From a certain perspective, Apple's visions of a computer revolution in 1984 presented itself as a political alternative. The fist Macintosh computer was built on ideas developed at Xerox PARC by Alan Kay and others in the 1970s. Xerox PARC was driven by free minds, user centeredness, and imaginary incentives, as a West Coast alternative to the conservative East Coast in the US history of computing. This was at a time when computing in the United States was primarily envisioned as a strategic tool.[10] So even though Apple's user-centered design that builds on studies of user behavior and cognition may seem mainstream and uncritical today, it is important to remember how it, along with other phenomena such as computer gaming, led the way to an alternative use of the computer.

In Scandinavia, the alternative and critical design tradition first presented itself as participatory design. Participatory design partly developed from the invention of object-oriented programming—a type of computer programming that builds on classes of objects on which to apply common methods. The Norwegian computer scientist Kristen Nygaard was one of the founders of object-oriented programing and participatory design (beginning in the 1960s with Simula, developed in league with Ole-Johan Dahl), and he collaborated with workers' unions in Scandinavia in the early 1980s around the development of new information technologies for the industries. Information technologies hugely affected labor processes, and Nygaard's aim was to develop systems that would not alienate the workers. In his collaboration, he found that the construction of a system's architecture and intent of the system always relies on a specific perspective on reality, which determines how the labor processes are labeled and broken down into classes as well as methods. To program is then not only a mathematical-technical process but also implies an understanding of the world that is negotiable—that is, an object-oriented interpretation of the world.[11]

In contrast to user-centered design that builds on an assessment *of* user behavior and cognition, 1980s' and 1990s' participatory design promotes the collaboration and design *with* the user. The model user is not just someone whose interest is user-friendliness but also is a skilled laborer who through a carefully guided design process can conduct research into their own professional tools. For instance, the Swedish informatics and design professor Pelle Ehn and Danish computer scientist Morten Kyng (who both played a role in founding Scandinavian participatory design) have outlined how planned design games can involve users in "mocking-up" a vision of a computer system that fits their reality. In a continuation of Nygaard's object-oriented approach, such mock-ups were used as templates for envisioned tools in a labor process, and often involved low-technological, moldable, and cheap materials such as cardboard boxes.[12] This method was, for instance, used in the UTOPIA Project, which focused on the development of information technologies in newspaper production. The print industries, and relation between, say, typographers and journalists, had been heavily affected by the introduction of digital typesetting, and there was an urgent need to involve the laborers in designing systems that reconfigured these relations in a meaningful and acceptable way.

Anthony Dunne and Fiona Raby also promoted the imaginary and critical alternative as a design strategy in what they termed "critical design." Unlike participatory design, the intention in critical design is not to create a de-alienating future design as a user-friendly tool and technological augmentation of the skilled laborer but instead to employ alienation strategically. Critical design is, as Dunne and Raby proclaim, "more of an attitude than anything else, a position rather than a method."[13] Neither a movement or an artistic practice, it is a design practice inspired by and seeking to employ the humanities' notion of critique as a way to conceptually and speculatively address a technological realm. The aim is "mainly to make us think. But also raising awareness, exposing assumptions, provoking action, sparking debate, even entertaining in an intellectual sort of way, like literature or film." As such, it is a design practice that is "not-just-design," and where the design research aims to produce objects that exceed their use value while conceptually addressing social, cultural, political, or other realities.[14]

Participatory design and critical design are two examples, among several, that demonstrate how the design of interfaces and field of HCI can perform critically.[15] Whereas the humanities traditionally perform critique analytically, and by way of reading and writing, participatory design and critical design highlight two distinct relational qualities of criticism by way of design. First, in participatory design, technical and social infrastructures are perceived as intrinsically related. Any technical reality is inherently a social reality, too, and collaborative and participatory development of tools, interfaces, technical workflows, and more is thus considered not only a means to user-friendly functionality but also a negotiation of existing hierarchies of power and control at the workplace.[16] Second, in critical design, concepts and realities are perceived as intrinsically related. Any conceptual understanding of a reality inherently gives rise to speculative and imagined realities as well.

Participatory design and critical design fundamentally differ, however. Susanne Bødker, a Danish professor in HCI (and one of the founders of Scandinavian Participatory Design as well as part of the UTOPIA Project in the 1980s), has identified a "third wave" in HCI that extends HCI's focus on the workplace to everyday life and culture. This wave includes Dunne and Raby's critical design, which she criticized for neglecting and even negating Scandinavian HCI's traditional commitment to the users: "The question is how the third wave could develop a productive, reflexive practice that makes more than artistic statements to provoke us." According to Bødker, a way to proceed "would be to move the co-determination framework outside the "factory gates."[17] The critical design implications of the ideas brought forward in this book in several ways seek to follow this challenge and make way for a different kind of interface criticism by design—an alternative third wave or potential fourth wave of HCI that seems ever the more relevant in a time of metainterfaces where the interface has turned into an industry and dominant actor in the tertiary sector, as elaborated in this book.

Yet this implies two fundamental shifts in perspective that potentially may bring participatory and artistic design practices closer. First, a shift away from the development of tools found in participatory design and toward the participatory design of an apparatus, and second, a shift away from the speculative realities found in critical design and toward a tactical formation of reality. Drawing on the insights of this book along with its focus on network and software culture and practices (and in particular, ideas of free software and tactical media) as well as two practical illustrations, the following will outline such a theory of an interface criticism by design.

The Tool

Ehn has argued that the introduction of computers into the workplace in the 1980s commonly marked a de-skilling of labor. This is the catalyst for participatory design, whose common purpose is to bring the laborers' crafting of tools back into the design of computer systems.[18] Put differently, in Ehn's Marxist approach the design of a tool is simultaneously the design of a labor process. But what is included and excluded in this labor process?

Ehn's line of thinking traces back to Douglas Engelbart's famous oN-Line System along with his ideas of "augmenting the human intellect" of the skilled laborer, but Ehn seems to specifically value the bodily aspect of the interface in this process. A tool, he writes, "is something that, when we have the skill to handle it, *is transparent to us; something that lets us have focal awareness on the task or on the material* we are working with. A good tool becomes an extension of our bodies."[19] In other words, the aim of the design process is to design an interface—through working with cognition, tangibility, and more—that allows the user to transcend the often-tedious and boring tasks of the human–computer interface itself. This ideal of what Ehn calls "conviviality" allows for the user's active engagement with the world—as if the tool was as enchanting and helpful as another human.[20] From this perspective, the design objective becomes the design of interactions rather than interfaces.

Participatory design thus both shares fundamental objectives with Apple's aforementioned user-centered design and yet differs. If participatory design in its efforts to design *with the users* has explicitly Marxist roots, user-centered design designs *for* the users, and is in this sense more oriented toward consumption. Though the means and ends differ, they both promote an ideal of seamless interaction. Although the skilled worker's tool may not appear convivial and human to an ordinary user, their shared aim remains: to keep the task at hand. In Apple's case, this implies a means to self-realization, which marks the beginning of a third wave of HCI where technology spreads from the workplace into everyday life and culture, and hence (today) the realm of what this book has labeled the metainterface. This shift from workplace into the everyday embraced new values of user experiences and meaning making, in favor of, for instance, efficiency, craft, and other values shared by Ehn and others.[21] The question of what these "improvements" in the natural history of interface design rely on, however, remains.

User-centered design and participatory design both represent a pragmatist view of the interface, in which the aims of interaction (the humane tool) supersedes the technological means. This pragmatism, seen in many areas of HCI and interaction design, displaces and represses the interface as a technological construct: it is regarded as an obstacle that stands between user and system.[22] Following the work of Winkler, such design traditions can therefore also be regarded as part of a wider history with different waves than Bødker's three-wave model, and former attempts to reappropriate presence and immediacy through technological means.[23] Hypertext and literary scholar Gregory Ulmer, for instance, asserts that Apple's user-centered design is an exclamation of how the "'twin peaks' of American ideology—realism and individualism—are built into the computing machine (the computer as institution)."[24] According to Ulmer, this is what drove the realism of US cinema, too, and led to "the 'invisible style' of Hollywood narrative films, and to the occultation of the production process in favor of a consumption of the product."[25] The "twin peaks" may open up for unexplored affordances of the networked computer, and revolutions such as the PC and graphical interface (i.e., individualism and realism).[26] When viewed from the perspective of interface criticism, though, such efforts to overcome the technological construct ignore the differentiation of representation that always takes place: the interface itself—on all levels—is a construct that embeds and performs particular values (ethics), assumptions (logics), and perceptions (aesthetics). Now more than ever, in a time of the metainterface when the interface seemingly disappears behind a veil of cultural consumption, social interactions, urban interaction, and other experiences as well as ways of sensing and making sense (which all seem to be for the common good), it seems relevant to discuss and address the role of the interface in the design of realities.

The Apparatus

Around the same time that Ehn and his colleagues defined the understanding of tools in participatory design, cinema studies developed a quite-different understanding of the media technology of cinema. In cinema studies, the role of critique, as Ulmer explains, has been to expose the suture (the seams rather than the seamlessness) that in turn expose the effects, "binding the spectator to the illusion of a complete reality."[27] On a more fundamental level, this affords an awareness of the cinematic apparatus itself. The notion of apparatus design as opposed to tool design offers new dimensions to participatory design that seem highly relevant in a time of metainterfaces.

Cinematic apparatus theory was established around 1970 as an understanding of cinema as a technological, institutional, and ideological machine. Professor of film studies Gertrud Koch points out that apparatus theory can be seen as a continuation of Walter Benjamin's aesthetics in which he considered the "technical equipment ... as the material basis of film aesthetics."[28] As discussed in detail in chapter 1, this critical understanding of the technical apparatus was central to Benjamin's materialist aesthetics, including the concept of tendency. Briefly put, the author as a producer (or the artist-filmmaker as a producer) explores the workings of the apparatus (its tendency) and critically reengineers it for new producers in order to, in Benjamin's words, "put an improved apparatus at their disposal."[29]

Benjamin argues that artists need to "enter into debate" with the apparatus instead of "thinking that they are in possession of an apparatus that in reality possesses them."[30] Following this perspective, the objective of this book has been to investigate how the interface relates to the commercial needs of globalized capitalism, territorial contestations in the urban, ways of making sense (of contemporary crises such as climate change and the war on terror), and even sense perception itself. Any attempt to put an approved apparatus at the disposal of others must begin with a realization of how the metainterface in reality also possesses us as users. On a fundamental level, the claim has been that contemporary cultural performativity, urbanities, and globalization all relate to the imperative logic of the metainterface: they are all wound up in a system of representation-computation, where all signs simultaneously function as signals in the networked computer, and all media are simultaneously instruments.

In order to understand how this system of representation-computation functions in the metainterface, and how it becomes integrated into perception, the artist and researcher Richard Wright discusses how interfaces create a differentiation by being algorithmically dependent on source data while creating a "perceptual independence from it." Commenting on how scientists working with fractal mathematics begin to see fractals everywhere in nature, Wright concludes, "Both algorithm and sensory vision are thus finally reunited in the cortex, in an endless circularity of computation and perception."[31] Like the mathematician seeing fractals everywhere, the system of representation-computation gradually becomes part of perception. In other words, the interface appears as if created for humans, but concurrently incorporates the human into a computational and algorithmic circulation system (or what in chapter 1 was referred to as a signal–computer interface), with deep impacts on the ways realities are sensed, perceived, formed, and handled.

Specifically for interface culture today, this incorporation happens on an abstract metalevel: the interface is in everything and everywhere, and yet this exchange of signals between computers is sealed off—apparently nowhere, in the cloud, out of human intervention. This displacement of materiality, which has become an intrinsic part of the metainterface, makes it particularly important to critically address the interface as apparatus in design rather than just the tool. In the spirit of Ehn and his coresearchers' focus on the users' codetermination, a material engagement with the apparatus, meaning an engagement with the technical apparatus and not just the role of a tool in a process of labor, is the kind of critical agency needed today.[32] A critique of the interface's hierarchies of control as an apparatus versus a tool implies that both the collaborative development of technical infrastructures (the kinds of critique performed by, for example, participatory design) and the conceptual speculation in other technological imaginaries (the kinds of critique performed by, say, critical design) should be viewed differently. Indeed, the whole assumption of being critical through practice should be viewed differently.

The Root to Change

The collaborative element of participatory design is essential to its criticism, but the collaboration and shared responsibility must be extended from the social to the technical apparatus. To write with a tendency, or become a producer as an author (in reference to Benjamin's seminal text), is a way of not only performing critical theory but also assuming responsibility for the institutionalization of critique and technology on which this happens (the apparatus). Benjamin even asks the writer to transform "from a supplier of the productive apparatus into an engineer who sees it as his task to adapt this apparatus to the purposes of the proletarian revolution."[33] For the author, this includes the responsibility for writing, printing, and publishing, yet in a larger technological perspective, authoring and engineering must be extended to how language and representation are connected to processes of computation in interfaces. Such a responsibility—as with any language—is naturally a shared and collective one.

Also in the tradition of participatory design, assuming responsibility for the design of interfaces is seen as a collective process. To Ehn and others, the focus has been on forming social relations from the outset of the collaborative design process. As Ehn and Kyng write, "The main point in design-by-doing using mock-ups is for everyone to get a hands-on experience, trying something new." This kind of "simulated future" ensures that once the participants have created a "mock-up," they will assume responsibility for its implementation: "almost any deviation to the final system in a mock-up requires some active work on behalf of the involved (future) users."[34] Mocking up a computer system with cardboard is in this sense an example of a "simulated future." It is a shared experience with something new that builds the community, and makes it assume a shared responsibility for implementing the change. The purpose of the mock-up is to create a design object that is not to be confused with the real system, but creates a common picture of the system. The implementation of this imagined system, however, not only depends on its functionality but also the organization in which the change occurs. In order to build stronger social ties between the people involved in the design process, for instance, Ehn and Kyng organized soccer games between typographers, journalists, and the designers themselves. This by no means implies that the implementation does not need to be negotiated (also known as the game of "class struggle," as they argue), but staging it as a design game may be a helpful tool.[35]

Ehn and Kyng represent a view on design that is quite similar to Jane Jacobs's and Richard Sennett's. As explained in chapter 3, to design for change in their view not only implies the design of an alternative reality but more so of a community that can handle change, too. The cultures that have taken root are the most agile and capable of dealing with change. To design critically, and also making way for critical alternatives, is in this perspective also a question of designing communities and recursive publics (as described by Christopher Kelty). In this respect, there are many overlaps between participatory design and this book's focus on the interface as an apparatus, where social reconfigurations are intrinsically related to material ones in the production system. There are important differences, though, to consider in a design situation.

Within participatory design, the context of the design community is perceived as consisting of social relations. The involved technologies are in this sense judged by their "convivial" capacities. That is, not only is the aim to humanize technologies; the involved technologies are (e.g., as exemplified in the cardboard mock-up) selected for their social affordance (cardboard does not necessitate technical skills and allows the participants to do fast mock-ups).

In contrast, a critical stress on the apparatus makes way for a different understanding of the design context. The context of the interface is intrinsically machinic, and not only consists of the social relations and practices with technologies but also the technological infrastructures themselves. Or put differently, if the critical and socially aware participatory design practices concentrate on how to make the best tools to support a desired social practice, then a design with a tendency will seek to bring forward the apparatus that binds the user to the tool.

According to Benjamin, the apparatus is designed with specific strategic functions that need to be reengineered (as described above). Also Michel Foucault, another important apparatus theoretician, has argued how the formation of an apparatus is a response to an urgent need, or as he proclaims, "The apparatus ... has a dominant strategic function." In the case of Foucault's own research into madness, sexual illness, and neurosis, the apparatus may have been "the assimilation of a floating population found to be burdensome for an essentially mercantilist economy."[36] The strategic imperative of the apparatus thus undertook the control of those "diseases." In the case of the design of computer systems at the workplace, similar interests may drive the assimilation of the user. Yet the point is not whether the management has a capital interest, or the laborers have an interest in making computer systems that are more humane; rather, it is to reflect how the system is driven by the logic of the apparatus itself. From this perspective, seeing the interface as a tool that seamlessly extends the human body seems to subjugate the technological apparatus.

As argued in this book, the subjugation of the technical apparatus is an intrinsic part of the metainterface, and in contemporary cultural platforms for computing, urban interface and cloud computing. Nevertheless, the contemporary workplace and language industry no longer seems to be as clearly identifiable as in the 1980s' newspaper industry. It instead seems as if the metainterface itself, as a cultural artifact that pervades all aspects of life, has become the locus of production (the workplace). Many others have sought to outline this contemporary class struggle of "the multitude."[37] The main objective of this book, however, has been to explore how this production system relates to sense perception. As seen, for instance, in the urban interface (chapter 3), the city is not just a context where the design of innovative urban interfaces opens up for new signification and practices (e.g., as Airbnb, changing an apartment into a bed and breakfast, or in other ways promoting the sharing of resources). The urban interface is also driven by an organizing principle—an optical and algorithmic unconscious—that defines it as a "not-just-human" apparatus and sensorium (as further examined in relation to the cloud in chapter 4). This nonhuman, unconscious, and subjugated perspective of the new masses of an interface culture is something that is feared and can be critiqued (e.g., as surveillance), but is also something that is maintained because it involves a profound desire for perspective.

Subsequently, assuming a common language in a design process is not merely a social construct that ensures that the alternative future can be developed and can take root. Users have an interest in the technologies themselves as well. They do not merely desire interfaces that match their needs for other activities (labor, social activities, etc.) but instead speak the language of the interface. Such an interest, following Winkler's work, not only buys into the imaginary new ways of living that allegedly come with the metainterface (e.g., a new emancipatory urban space) but follows a desire to speak and share a language as an aim in itself, too. The metainterface also spreads because people have a desire for new perspectives, but they need to critically see this perspective in order to speak freely. This points to the aesthetic qualities of the language of the interface, which is not its beauty but rather its ability to point to a condition in language in which correctness and authority remain undefined (as many of the examples in this book demonstrate). Or as the French philosopher Jacques Rancière writes, the literary (as a way of writing) is that which "circulate[s] without a legitimating system defining the relations between the word's emitter and receiver."[38]

"Not-just-art"

Assuming the right to speak naturally involves an element of expressivity (as also asserted in chapter 3). In this sense, there is a strong connection between critical design and a will to bring forward the apparatus that binds the user to the tool (bringing forward the tendency of the interface). Yet here, too, there are substantial differences that relate to the tactical dimensions of critical interface design (and the streets and underbelly of digital culture), and bypass the nonproductive distinctions between art and not art.

In his critique of the transparent interactivity that characterizes most design in the workplace and elsewhere, Dunne quotes Paul Virilio, who contends that user-friendliness is "just a metaphor for the subtle enslavement of the human being to 'intelligent' machines."[39] Dunne continues, noting that "this enslavement is not … to machines, nor to the people who build and own them, but to the conceptual models, values and systems of thought the machines embody." To address these conceptual models or apparatuses, Dunne proposes to "poeticize" the relation between people and electronic objects. To poeticize means to construct interactivity that operates in the land between transparency and opaqueness, or what he labels "the pet" and "the alien": "the pet offers familiarity, affection submission, and intimacy, the alien is the pet's opposite, misunderstood, and ostracized." Dunne later remarks, "If aliens and user-unfriendliness are to be the alternatives to pets and user-friendliness, this user-unfriendliness does not have to mean user-hostility. Constructive user-unfriendliness already exists in poetry."[40] Poetry, in other words, is an example of something that operates in the area beyond the passing of information. In certain formalist traditions, it can be used to "counteract the familiarization encouraged by routine modes of perception."[41]

Though referring to poetry, Bertolt Brecht's alienation of the audience in theater, and also numerous art installations, Dunne later exclaimed (with Raby) that critical design "is definitely not art."[42] In its experimental nature and institutional critique, artistic practice is most often prevented from becoming a sustainable practice, and is rarely seen as having any function outside the art world. Thus, there is a need for a "functional estrangement," or "conditions where users can be provoked to reflect on their everyday experience of electronic objects." This involves an aesthetics of use that is "grounded in functionality." The conceptual models, values, and systems of thought that electronic objects embody can be critically challenged by "gadgets" or "philosophical toys" that self-consciously contest the usually implicit concepts of design.[43]

In several ways, Dunne's position and examples resemble the self-perception of software art brought forward around the same time as Dunne's seminal book *Hertzian Tales* (first published in 1999). In his essay "A Means of Mutation: *Notes on I/O/D 4: The Web Stalker*" (from 1998), Matthew Fuller argues for the need to bypass the unproductive dichotomy of art and nonart, or culture versus counterculture. Fuller's own collaborative work *Web Stalker* (figure 5.1) is an illustration of how this may be done. *Web Stalker* is a functional browser that instead of showing the web as pages, scans for and visualizes the link structure of the World Wide Web and displays the HTML code. It is, as Fuller labels it, an example of "not-just-art," suggesting the possibility of cultural forms that transgress the dichotomy of art and functional design. By being not just itself, and being used, it can be assimilated into "possible circuits of distribution and effect [that] in this case means something approaching a media strategy."[44]

The poetic design of *Web Stalker*, however, is quite different from the industrial designs of alienating functionality put forward by Dunne. Dunne, for instance, quotes Ettore Sottsass, who in the late 1950s designed the Elea 9003 computer for Olivetti:

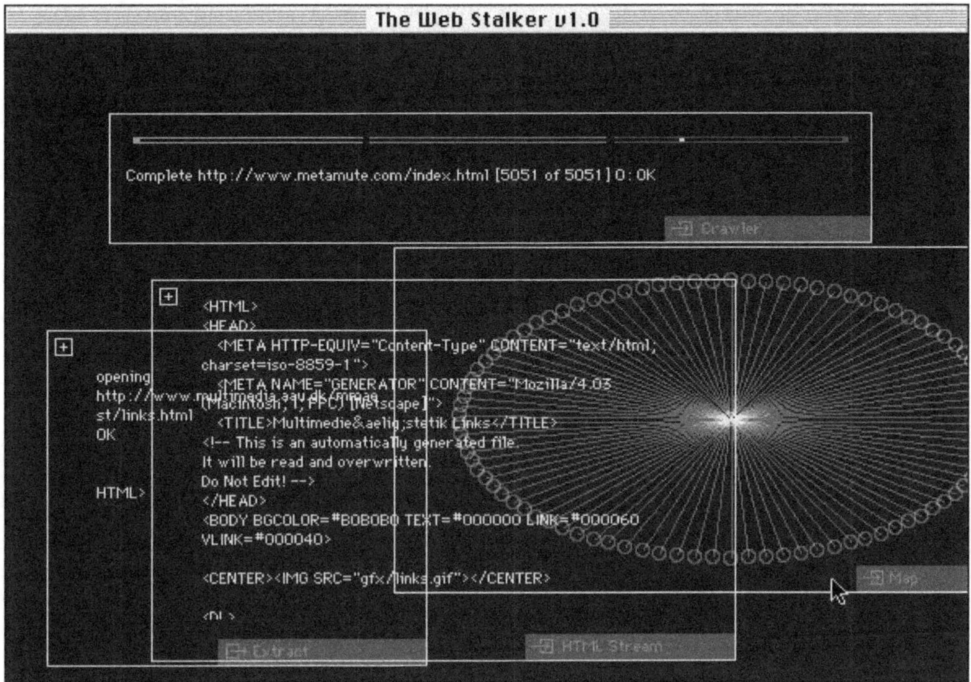

Figure 5.1
Web Stalker (1997) by I/O/D. A browser that instead of showing the World Wide Web as pages, scans for and visualizes its link structure, and displays the HTML code. Courtesy of I/O/D. Screen capture of browser window.

One ends up immediately designing the working environment; that is, one ends up conditioning the man who is working, not only his direct physical relationship with the instrument, but also his very much larger and more penetrating relationship with the whole act of work and the complex mechanisms of physical culture and psychic actions and reactions with the environment in which he works, the conditionings, the liberty, the destruction, exhaustion and death.[45]

Dunne suggests that the quality of a poetic design lies in its challenge of user-friendly designs that reduce "the relationship between people and technology to a level of cognitive clarity."[46] Fuller, in contrast, emphasizes that *Web Stalker's* visualization of nodes and links in the network is more than a critical statement and provocative critique of the conventional browser's interface. As he asserts, "The web stalker works as a kind of 'tactical software' but it is also deeply implicated within another kind of tacticity—the developing street knowledge of the nets."[47] In other words, in contrast to Dunne's critical design, not-just-art builds on net culture and is explicitly tactical.

Tactical media was a term used at the time by David Garcia and Geert Lovink to coin the forms of interventionist media activism that flourished with the cheap, DIY media of consumer electronics (and especially involving the Internet as a media network) by those who felt "exploited by groups and individuals who [felt] aggrieved by or excluded from the wider culture."[48] The term itself derives from Michel de Certeau's *The Practice of Everyday Life*, in which he foregrounds the use of representations. In terms of media, what is at stake is the question of "how we as consumers use the texts and artifacts that surround us."[49] The answer is "tactically," or as "tactical media." In this sense, Fuller's proposal of not-just-art is a way of suggesting that the strategy of a ruling power, which binds the user to the tool in particular ways (through the design of computer interfaces and browsers), can be countered. He points to the Internet itself as a new strategy, not by tearing the clarity of the browser tool down, but by offering another tool that derives from the tactics of the streets. Put differently, it is a tactic that depends on, for instance, the construction of nodes in the network by a DIY net culture, which sets up web servers and creates new kinds of sociotechnical apparatuses (as also described in relation to the vernacular web in chapter 4). When Fuller stresses the strong ties between not-just-art and free software, it is exactly for this reason: it is a work of not-just-art derived from a new and networked mode of production of a recursive public.[50] With the terms suggested by this book, *Web Stalker* is not just an example of poetic design but instead the author as a producer, which in Benjamin's thinking, means an author who reflects the means and conditions of production itself as a tendency.

In retrospective, *Web Stalker*'s notion of how the Internet is vernacular—how it is networked, and the potential circulator of freeware and nonformulaic software—has almost proven to anticipate much of contemporary program design seen on the platforms of the interface industry and social media. Today, visualizations of networks is, for instance, a feature in what is commonly conceptualized as a "social graph," which is not really a graph but rather an illustration of the relations between users, places, and other things they interact with online. The Internet no longer consists of URL addresses connected in a network (as illustrated in *Web Stalker*); it is composed of phenomena like Facebook's Open Graph platform that allow for, say, combining data from Spotify with Facebook's network of friends (as a user service, everything people consume, produce, or like will eventually end up on somebody's News Feed on Facebook). The social graph may look similar to *Web Stalker*, but in its mode of production (as a metainterface), it is substantially different. As indicated throughout this book, the alternative strategy of the network—and even its visions of new ways of living, producing, and sharing—have become platformed, territorial, and clouded. In this sense, there is a need for *interface tactics* that take into consideration the whole sociotechnical apparatus of the cultural, urban, and cloud interface.

As a concluding statement, this book ends by describing two cases of interface criticism by design, conducted collaboratively by the authors of this book together with a number of other people. Each in their own way, the cases address (topically for this book) the production as well as poetics of writing and language.

Case 1: *The Poetry Machine*

Is the printed book (codex) disappearing, and are digital formats such as the Internet or e-book superseding it? This has been discussed for decades, and as pointed out in chapter 2, digitization influences reading and writing, but literary institutions such as libraries are being challenged, too. In *The Late Age of Print*, Ted Striphas maintains that the printed book is not disappearing but instead that "books exist in a more densely mediated landscape than ever before." He focuses on book publishing, and shows that the problematic relations between the book industry and libraries are not new. These relations have intensified with e-books, which are often not owned but rather licensed and controlled even after purchase. This of course severely limits the possibility of loans both privately and publicly in libraries. Striphas argues that e-books have an important story to tell about the "logic of capitalist accumulation," and "more than fifty years' worth of effort to render problematic people's accumulation and circulation of printed books."[51]

The Poetry Machine (figure 5.2) was developed in 2012 as a way for libraries to exhibit electronic literature.[52] The installation consists of three sensor-equipped books through which (up to) three simultaneous users can compose poems on a screen, and then get them printed on small receipts and stored on a website. When seizing a book, the user is assigned a sentence from this book out of approximately a hundred different sentences. Each sentence exists in three variations, which the user can choose to drag into the writing space. After a limit (e.g., 350 characters) is reached, by combining the books and sentences, the poem is finished, printed, and stored online.

The Poetry Machine was designed as a collaborative project between librarians, authors, and researchers, and the design has focused on critically addressing the digitization of literary culture—that is, on the tendency of the literary apparatus. *The Poetry Machine* allows users to experience digitization through the composition of poems and interaction with the installation. In this way the installation seeks to make the apparatus of digitization sensible. Apart from making a usable and meaningful literary installation, it proposes that digitization does not just make the book disappear into virtual

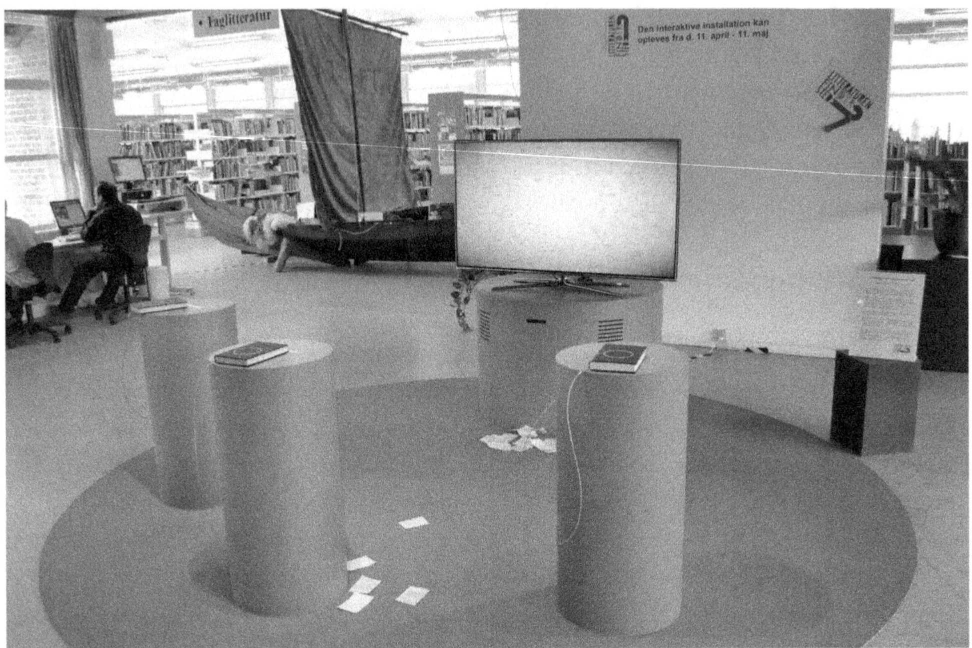

Figure 5.2

The Poetry Machine (2012–) by Woetmann, Aarhus University/CAVI, and Roskilde Library. The installation consists of three sensor-equipped books through which (up to) three simultaneous users can compose poems on a screen, get them printed on small receipts and stored on a website. Courtesy of Roskilde Library, Aarhus University, and CAVI. Photo by Søren Pold.

libraries but instead on a more fundamental level changes writing itself. Furthermore, it suggests a tactic to comprehend and act against the disappearance ("burning") of the book that has happened at many libraries. As not-just-art, it explores how tactics from electronic literature challenge traditional literary understanding through three interconnected levels.

First, because the user cannot open and read the installation's closed and bolted books in a traditional fashion, it brings with it a reflection on the status of the book. The books have to be "read" through interacting with the screen. More than the disappearing of the book, however, this points to how books, print, and digital text coexist as technologically reproduced media. The installation stages an interaction with books, thereby producing text on a screen that ends on a printed receipt and website, and pointing to the contemporary conditions of literary, textual media. Currently, there is a strong interest in what has been labeled the "aesthetics of bookishness"—a renewed interest in understanding the materiality of the book and reexploring what it becomes after digitization (that is, in the "post-digital" era).[53] Adding to these discussions are the ways that new limitations, monitoring, and forms of control are part of the meta-interface industry, including its literary media such as e-books and web platforms (as described in chapter 2). *The Poetry Machine* was designed as a hybrid alternative to the immaterial e-book, and has produced both a printed poem collection and e-book.[54] In this sense, the immediate opposition between print and digital becomes a question of reflective relations between different material layers of production and consumption, including their different qualities and problems. The reader of the installation can navigate between different literary media, from book to screen and back to print, and is invited to reflect on the differences, exploring relations between material dimensions of the changing media and their semiosis.

Second, on the textual level, *The Poetry Machine* can be characterized as a platform for what Kenneth Goldsmith has described as "uncreative writing," where users make attempts at creating a syntagmatic sequence, a poem, from an existing paradigmatic landscape of sentences.[55] Peter-Clement Woetmann's written text (the participating author) prepares for different kinds of relations between the sentences on the metaphoric dimension (e.g., the different themes of the installation's three books and how they influence the creation of poems) and the metonymic one (e.g., relations that emerge between sentences, and the way that they comment on the situation, experience, and media of the installation). Besides, there is a strong phatic dimension related to the many personal pronouns, which reflects both the "I" and "you" of the interaction and reading, authorship and attribution, and relation between human and machine text, or between the writerly "I" and machinic "you."[56]

Third, both the hybrid, material reflection and uncreative textual writing are important parts of an interface that is also performative and social. The interaction, reading, and writing is a social and performative activity (which can take anywhere from minutes to hours), in which users can watch other users, learn from them, and collaborate. Some users, experience something about how interfaces are constructed and modulated: how the different layers of the interface reflect on each other. As evaluation reports have shown, many of the users reflect on authorship, the ways digital text is composed, and how algorithms and corporations such as Google and Facebook control that digital text.[57] Even if most of the users were unaware of electronic literature as an art form, many of them reflected on the power balance and control of contemporary textual interfaces, and how this balance is often hidden behind user-friendly metaphors.

Manuel Portela argues that "programmed performativity creates a metareading position" where the metareader ultimately tries to read both the produced text and algorithmic interface that produces this text.[58] Related to *The Poetry Machine*, this kind of metareading might lead to moments where media and interface resonate with the text in more or less meaningful ways, echoing the poetic design as a kind of critical design. In other words, it is when the textual output and interface interrelate that critical as well as self-reflexive interpretation emerges. This obviously does not happen without an active, interpreting user: the critique only happens through extended poetic coreading and cowriting.

The Poetry Machine can be seen as an attempt at interface criticism by design. It is designed to formally interrogate the tendency of the digitalization of literature and the library. It questions the material conditions of the printed book versus digital textual forms along with the process of reading and writing, including the authorial function, and the way meaning is generated, spread, and harvested in a social and collaborative space. Even though *The Poetry Machine* is definitely designed from a critical beginning, it is not designed with a specific critical point in view but rather as a way to open up space for interpretation through exploring the tendency of literary digitalization. It sketches out alternatives to the way the Internet and hypertext has become instrumentalized as an interface industry, such as by allowing the users to compose and control their text rather than harvesting the text in order to profile the user.[59] As an alternative to the platforms of the interface industry (e.g., the e-book), *The Poetry Machine* is designed for the library as a social, material, and institutional space. It aims to show that digitalization is more than e-books and databases, and literary culture in the era of the interface industry needs open, cultural spaces like libraries in order to reflect on current digital developments. At a time where many libraries organize makerspaces and design workshops to keep pace with the demands of knowledge dissemination in the digital age, *The Poetry Machine* instead points to how electronic literature already "knows" about its digitalization (as tendency).

Case 2: *A Peer Reviewed Journal About_*

As also argued by Bernard Stiegler in his critique of Theodor Adorno and Max Horkheimer's "The Culture Industry: Enlightenment as Mass Deception," media criticism should not only understand how new media are objects of thought and perception but also how critique itself—through a technological apparatus that includes academia as well as book production—is part of perception and memory. Adorno and Horkheimer do not, according to Stiegler, reflect that "the very possibility of 'culture,' and thus of 'spirit,' relies on technics."[60] Stiegler's critique is further explained in this book's chapter 2, but a challenge remains: How to perform critical writing critically? As a continuation of this, the questions are, more specifically, How can critique be informed by practices from network culture, and how can the practices of network culture themselves potentially form a recognizable academic field? Such questions are deeply wound up in the practices and economies around academia.

As also discussed in chapter 2, the mechanical reproduction of books is historically related to the development of conveyor belt capitalism.[61] Hence, critique itself—which follows Gutenberg's press, and grows out of a necessity to asses, review, and discuss books—cannot be separated from the production apparatus of the book. The printing press brings with it an institutionalization of critique as a specialized commentary that is separated from other forms of practice. There is, in other words, a historical tradition of performing critique in writing, which also displaces critique as something that is practiced. How can one free critique from this institution of writing, and look for alternatives that do not necessarily explain the apparatus but instead write with it, as a practice? This seems to be the inherent question of Stiegler's argument against Adorno and Horkheimer.

The publication platforms *A Peer-Reviewed Newspaper* and *A Peer-Reviewed Journal About_* (*APRJA*) (figure 5.3) are attempts to imagine and form a research community with built-in publication platforms that critically reflect the apparatus of critique and research.[62] The publications are structured around a process that begins with an open call to practice-based and academic PhD researchers on mailing lists such as spectre and nettime. The call is an invitation to a workshop that addresses a timely topic with particular currencies in network cultures, and is typically developed in collaboration with transmediale, matching the festival's thematic framework. On review and acceptance of applications, the participants publish their preliminary writings online (on a blog), followed by comments. At a workshop—hosted by varying partner institutions—participants meet, present and discuss their work, and collaboratively produce a newspaper publication.[63] The newspaper is presented at the transmediale festival as well as distributed through other personal channels. Following the newspaper, all participants are invited to submit lengthier presentations, which undergo double-blind peer reviews for the online, open-access journal *APRJA*. A number of aspects appear relevant in relation to the design of a critical platform for research publication.

First, the newspaper and journal explicitly change the status of printed critique and research. What matters are not only online open access and gratuity but also the imagination of research being distributed physically and "freely" through communities that meet at locations such as the transmediale festival.

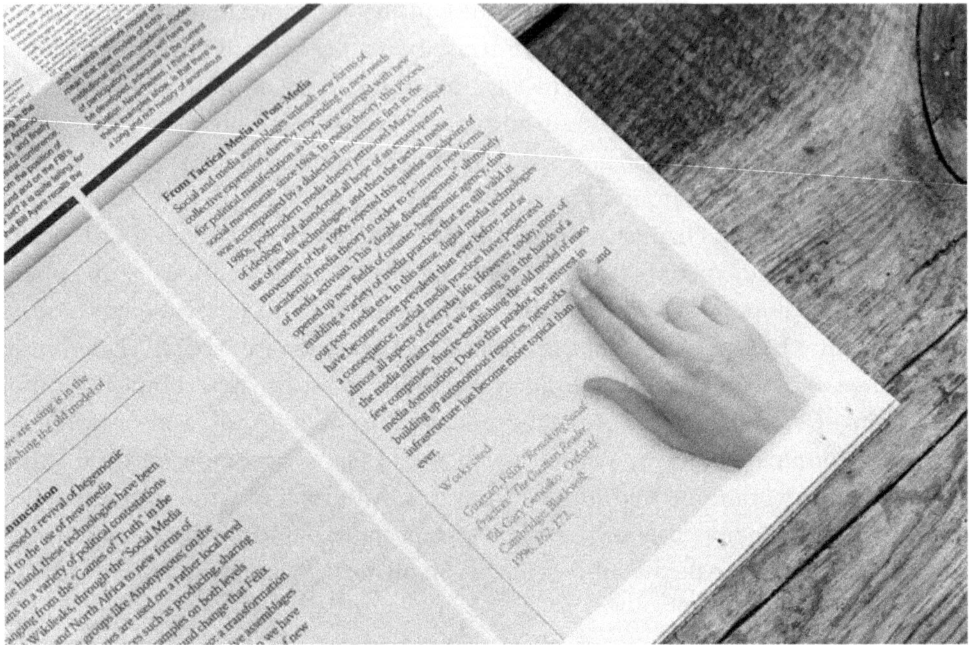

Figure 5.3
*A Peer-Reviewed Journal About_* (2011–) by Christian Ulrik Andersen and Geoff Cox in collaboration with transmediale festival and numerous partners. Photo of the newspaper issue "Researching BWPWAP!" designed by the Laboratory of Manuel Bürger. The publication platforms *A Peer-Reviewed Newspaper* and *A Peer-Reviewed Journal About_* are attempts to imagine and form a research community with inbuilt publication platforms that critically reflect the apparatus of critique and research. Courtesy of APRJA and its contributors. Photo by Manuel Bürger.

Second, the production process explicitly seeks integration into the existing apparatus of academic publication. The double-blind peer review of the journal allows for bibliometric measurements. Such measurements are important in the public funding of universities, but also crucial in building personal curricula that are valued by the universities. In short, most academic posts and tenure positions to a large extent depend on the applicant's list of peer-reviewed publications.

Third, the platforms seek to replace the academic network of reviews, quotations, and so on, by the physical processes of communication, discussion, and collaboration. The precarious situation of both research institutions and individual researchers largely builds on austerity measures of peer-reviewed articles and quotations. In a response to this fetishization of the network, the peer reviews and references are no longer abstract but concrete, and build on human relations and collectives that potentially may take root and handle change (as envisioned by Sennett and explained in chapter 3).

The "free model" of the project (the first characteristic) must be seen in light of how academia has been and could be disrupted by network culture. Compared to other areas of cultural production, the academic press appears (on the surface) strangely untouched by the digitization of print. Unlike publishers of music, for instance, which have largely been replaced by first pirates and then software and hardware companies, academic publishing seems to be flourishing.[64] Academic publishing is for many companies (and mostly because of expensive university library subscriptions) a good business that seemingly has remained unaffected by the disruptive innovation of digital writing and copying. Even open access to journals is incorporated into the business of publishing, and in the spirit of openness.

As explored in chapter 3, openness itself—though intrinsically related to the new production paradigm of digitization—also opposes a perception of the apparatus as something fundamentally "free" and shared between its participants. In other words, it tends to reduce the importance of sharing the apparatus of research (and research publication) to questions of business and free (gratis) access. Arguably, gratis access to knowledge is greatly appreciated by the people who remain outside the academic institution, such as artists and hackers who are actively engaged in network culture and events such as the transmediale festival. Nevertheless, to them the question of building an apparatus does not easily explain itself in economic terms as a question of gratuity.

In the mid-1990s, the network culture suggested other alternatives. Network culture can (as an oversimplification) be characterized as basically anti-institutional and therefore having a built-in skepticism toward academic institutions, including publication channels and accreditation systems. Geert Lovink, the founder of the Amsterdam-based Institute of Network Cultures, was one of the most prominent people to express this anxiety. In 1995, in the early days of the World Wide Web, he (together with Pit Schultz) formed a mailing list called nettime, which has since then formed numerous sublists. Nettime has featured seminal productions, essays, and discussions (many of them quoted in this book), and was the prime publication channel for critique and research into network culture. Although there are many other channels today, its status still remains (now run by Felix Stalder and Ted Byfield). Many academic fields since then have created mailing lists, but nettime, as a moderated and nonhierarchical discussion board, differs substantially from traditional academic organizational structures (with conferences, journals, associations, boards, presidents, etc.). It is the sort of apparatus that fits the recursive public of the free network of peers that cares for and tries to shape its own means of existence.

The free model that the *APRJA* builds on explicitly also relates to the precarious situation of the university (the second characteristic). Research is not easily detached from its institution, the university. In order to survive, the network may at times need the recognition and accreditation of a system that sustains many of its participants. One may argue, too, that there are certain advantages of the ancient procedures of academic institutions. If all goes well, peer review is a system that ensures that reflections and critique are less opinion driven, that they are not (just) driven by fancy new concepts but instead relate to, respect, and quote prior conceptualizations, and that they are less repetitive and more rigorous in their documentation and argumentation.

The overall project, however, must also be understood in light of a how contemporary academic research and publication has become wound up in an all-encompassing network (the third characteristic). In research, there seems to be no reality of research outside the network of impacts and quotations that have become markers of excellence (maintaining many institutions and academics in precarious situations). As a critique of the network, one may also assert, as does Nishant Shah, cofounder of the Centre for Internet and Society in Bangalore, that "the network is an opaque metaphor, conflating description and explanation. So it becomes the object to be studied, the originary context that produces itself, and the explanatory framework that accounts for itself."[65] In other words, in a network society—where every social activity is seen through the lens of the network—networks become a self-fulfilling new ontology for the human being. There is no being without it being networked. As actors in a network that generate networks (relations, transactions, traffic, archives of traces, etc.), there is no longer a living human being who is identifiable as part of a social or political community.

Shah's critique is relentless, and of course directed toward social media, but his plea to create a human ontology outside the precariousness of the computational networks of the metainterface (and its signal–computer interface) should be taken seriously, including in relation to academia and academic publishing. The university, which is increasingly seen and governed through the lens of networks (or "excellence" measured in quotations, or other network impacts), needs living beings and communities whose existences as researchers and research communities are not just dependent on their visibility in a network of austerity measures. As discussed in chapter 1, the networks of the metainterface enforce a machinic temporality, meaning that only the networked have a future. In opposition to this, the newspaper and journal seek to stage situations in which participants are allowed to envision, form, and share other futures.

Designing Critique

To investigate and design the apparatus rather than the tool, the seams rather than the seamlessness, and study how contemporary culture is stitched together by the metainterface, is in this book's perspective an investigation of and design with the interface's tendency. It is at once an exploration of a technological apparatus and how this apparatus affects sense perception (meaning not just what is sensed but also how it is sensed). This dual objective has been prevalent throughout the book and its investigation of how metainterfaces are built on the production as well as consumption of language, and how this alters the perception of the self. The dual objective has examined how interfaces take part in the semiotization of the urban, how this changes the perception of time and space, how interfaces are applied in cloud computing, and how this virtualization of a technical infrastructure alters the perception of a globalized world and its various crises (such as the war on terror and climate change). A reasonable conclusion to these investigations is that sense perception operates through a technological unconscious—through hidden machinic languages, perspectives, and infrastructures. The technical apparatus of the interface, however, easily escapes attention; it is black boxed, hidden away, and in constant disappearance as an unconscious. So how does one reveal and reengineer this tendency (the workings of the apparatus) in practice in order to put an improved apparatus at the disposal of others?[66]

The two cases illustrate how to design with a tendency, and how to engage with the apparatus of reading and writing text. Textual production and consumption is not just a question of using a digital tool that affords new creative and social possibilities; rather, it is an apparatus that encompasses the printing press, the book and its digitization, the library, academia, and more, and inscribes readers, writers, and machineries into a larger system of production. The engagement with this apparatus is not just art but also a way to actively and tactically put an improved apparatus at the disposal of others. As much as anything, this is not simply a question of designing an apparatus for the other (as an exchange of product) but instead engaging the other in a process that aims at sharing the apparatus between participants and building communities (rather than just networks for the exchange of commodities).

In this, there is still much to be learned from prior experiments with participatory design. Yet the kind of productivity involved in the process also opposes the pragmatism of tool development and mobilizes quite-different energies. At first sight, and as noted by Andy Goffey, computing—with its logical formal-material calculus—"seems to carry to an extreme an understanding of technology as a utilitarian tool predicated on a reframing of the world as a set of calculable quantities."[67] The two projects—along with the many examples in this book—have hopefully demonstrated an interesting imprudence and obstinacy toward the pragmatism of tools that seemingly comes with computing. This relates explicitly to both free speech (as in free software) and the affective energies of pleasures that relate to the free speech.

The book's investigations have been through critical analysis and theory. In a challenge to the ancient understanding of critical theory as a mode of contemplation—looking at the world, analyzing it, and asking questions to gain insight—the book simultaneously proposes that interface critique is a practice.[68] Interface criticism, when practiced, does not merely explain things but instead humorously points to the inner workings and mechanisms of the apparatus. From this perspective, it is a release of the technological unconscious that also makes way for the free, tactical, and participatory reengineering of our technological shared future.

Notes

Introduction

1. Geert Lovink, *Zero Comments: Blogging and Critical Internet Culture* (New York: Routledge, 2008), 44.

2. Gilbert Seldes, *The Seven Lively Arts* (New York: Harper and Brothers, 1924).

3. See also David M. Berry and Michael Dieter, *Postdigital Aesthetics: Art, Computation, and Design* (Basingstoke, UK: Palgrave Macmillan, 2015).

4. In his work on the philosophy of contemporary art, Peter Osborne also refers to the art that reflects the production of contemporaneity as postconceptual art, though he does not explicitly discuss artistic production in relation to technological networks. Peter Osborne, "The Postconceptual Condition," *Radical Philosophy* 184 (2014): 19–27.

5. Joseph L. Bower and Clayton M. Christensen, "Disruptive Technologies: Catching the Wave," *Harvard Business Review* 73, no. 1 (1995): 43–53; Joseph A. Schumpeter, *Capitalism, Socialism, and Democracy* (New York: Harper and Brothers, 1942).

6. Christophe Bruno, *Fascinum*, 2001, http://www.unbehagen.com/fascinum/.

7. See http://www.tenbyten.org/press.html.

8. Christophe Bruno, "Collective Hallucination and Capitalism 2.0: Scale-Free Elections in France," in *Interface Criticism—Aesthetics beyond Buttons*, ed. Christian Ulrik Andersen and Søren Pold (Aarhus: Aarhus University Press, 2011), 279–292.

9. *Artwar(e)*, 2010–2012, http://www.artwar-e.biz/en/home/.

10. See http://www.artwar-e.biz/en/analyses/hype/1/.

11. César Escudero Andaluz, *Inter_fight*, 2015, https://escuderoandaluz.com/2015/06/08/inter_fight/.

12. Matthew G. Kirschenbaum, *Mechanisms: New Media and the Forensic Imagination* (Cambridge, MA: MIT Press, 2008).

13. Naturally, the notion of archaeology also refers to the media archaeologies presented by Erkki Huhtamo, Jussi Parikka, Wolfgang Ernst, and others. Erkki Huhtamo and Jussi Parikka, eds., *Media Archaeology: Approaches, Applications, and Implications* (Berkeley: University of California Press, 2011).

14. Tung-Hui Hu, *A Prehistory of the Cloud* (Cambridge, MA: MIT Press, 2015).

15. Joana Moll, *CO2GLE*, 2015, http://www.janavirgin.com/CO2/. For a more detailed description, see www.janavirgin.com/CO2/CO2GLE_about.html.

16. Chien A. Chan, André F. Gygax, Elaine Wong, Christopher A. Leckie, Ampalavanapillai Nirmalathas, and Daniel C. Kilper, "Methodologies for Assessing the Use-Phase Power Consumption and Greenhouse Gas Emissions of Telecommunications Network Services," *Environmental Science and Technology* 47, no. 1 (2013): 485–492.

17. In their account of ubiquitous computing twenty years after Marc Weiser's seminal work, Paul Dourish and Genevieve Bell also discuss how Weiser's future vision compares to a much more messy, disharmonic, and perceptible technological reality: "Computational technologies are embedded in social structures and cultural scripts of many sorts; ubicomp technologies prove also to be sites of social engagement, generational conflict, domestic regulation, religious practice, state surveillance, civic protest, romantic encounters, office politics, artistic expression, and more." Paul Dourish and Genevieve Bell, *Divining a Digital Future: Mess and Mythology in Ubiquitous Computing* (Cambridge, MA: MIT Press, 2011), 41–42.

18. Benjamin Bratton has described the interface industry as a "Cloud Layer." Benjamin H. Bratton, *The Stack: On Software and Sovereignty* (Cambridge, MA: MIT Press, 2015).

19. Matthew Fuller, "A Means of Mutation: Notes on I/O/D 4: The Web Stalker," in *Behind the Blip: Essays on the Culture of Software* (New York: Autonomedia, 2003), 62.

Chapter 1

1. Espen J. Aarseth, *Cybertext: Perspectives on Ergodic Literature* (Baltimore: Johns Hopkins University Press, 1997).

2. See, for instance, George P. Landow, *Hypertext: The Convergence of Contemporary Critical Theory and Technology* (Baltimore: Johns Hopkins University Press, 1992).

3. Aarseth, *Cybertext*, 1.

4. Ibid., 3.

5. Jay David Bolter and Richard Grusin, *Remediation: Understanding New Media* (Cambridge, MA: MIT Press, 1999).

6. Lev Manovich, *The Language of New Media* (Cambridge, MA: MIT Press, 2001), 69.

7. Matthew Fuller, ed., *Software Studies: A Lexicon* (Cambridge, MA: MIT Press, 2008).

8. Christian Ulrik Andersen and Søren Pold, eds., *Interface Criticism: Aesthetics beyond Buttons* (Aarhus: Aarhus University Press, 2011).

9. Alexander R. Galloway, *The Interface Effect* (Cambridge, UK: Polity Press, 2012); Lori Emerson, *Reading Writing Interfaces: From the Digital to the Bookbound* (Minneapolis: University of Minnesota Press, 2014); Branden Hookway, *Interface: A Genealogy of Mediation and Control* (Cambridge, MA: MIT Press, 2014).

10. Benjamin H. Bratton, *The Stack: On Software and Sovereignty* (Cambridge, MA: MIT Press, 2015). Bratton in several ways portrays a technical landscape similar to the one in this book, yet without having the same focus on art and aesthetics.

11. Wendy Hui Kyong Chun, *Programmed Visions: Software and Memory* (Cambridge, MA: MIT Press, 2011).

12. N. Katherine Hayles, *Writing Machines* (Cambridge, MA: MIT Press, 2002); Manuel Portela, *Scripting Reading Motions: The Codex and the Computer as Self-Reflexive Machines* (Cambridge, MA: MIT Press, 2013). Benjamin's materialist perspective on writing will be addressed in more depth later in this chapter.

13. See, for instance, Régis Debray's materialist criticism of semiology: "We are presented with a manufactured knowledge whose basis and backdrop are missing the concrete networks of knowledge (labs, scientific community, procedures, congresses, journals, etc.), whose Reason lacks a sequence of inscriptions, and whose relationship of meaning has no relations of force. A supreme fiction of the work without tools or workers?" Régis Debray, *Media Manifestos: On the Technological Transmission of Cultural Forms* (London: Verso, 1996), 72.

14. Florian Cramer, *Anti-Media: Ephemera on Speculative Arts* (Rotterdam: nai010 publishers, 2013), 68.

15. According to David Berry, object-oriented ontology and speculative realism "reflect a worrying spirit of conservatism within philosophy" that "discount the work of human activity." David M. Berry, "The Uses of Object-Oriented Ontology," May 25, 2012, http://stunlaw.blogspot.dk/2012/05/uses-of-object-oriented-ontology.html.

16. Consequently, the book not only follows on from former works on interface criticism but also from Jussi Parikka's call for a multiplicity of materialisms. Parikka argues against some contemporary forms of materialism's limited focus on objects as nonreducible to signification, and argues for the need to consider how media materials are interwoven with signs, and how they are also nonsolids and intertwined in processes (i.e., the kind of materiality that is inherent to technical media). Jussi Parikka, "New Materialism as Media Theory: Medianatures and Dirty Matter," *Communication and Critical/Cultural Studies* 9, no. 1 (2012): 95–100.

17. Bruno Latour and Peter Weibel, *Making Things Public* (Cambridge, MA: MIT Press, 2005); Bruno Latour, "Why Has Critique Run out of Steam? From Matters of Fact to Matters of Concern," *Critical Inquiry* 30, no. 2 (2004): 225–248; Bruno Latour, "What Is the Style of Matters of Concern" (Assen: Van Gorcum, 2008).

For a critique of Latour, see, for example, Benjamin Noys, "The Discreet Charm of Bruno Latour," in *(Mis)Readings of Marx in Continental Philosophy*, ed. Jernej Habjan and Jessica Whyte (London: Palgrave Macmillan, 2014), 195–210.

18. "The moral is simple: only partial perspective promises objective vision. All Western cultural narratives about objectivity are allegories of the ideologies governing the relations of what we call mind and body, distance and responsibility. Feminist objectivity is about limited location and situated knowledge, not about transcendence and splitting of subject and object." Donna Haraway, "Situated Knowledges: The Science Question in Feminism and the Privilege of Partial Perspective," *Feminist Studies* 14, no. 3 (1988): 595, 583.

19. Peter Bøgh Andersen, "Vector Spaces as the Basic Component of Interactive Systems: Towards a Computer Semiotics," in *VENUS* (Aarhus: Aarhus University, 1991), 20.

20. Frieder Nake, "Human–Computer Interaction: Signs and Signals Interfacing," *Languages of Design* 2 (1994): 204.

21. Ibid.

22. Ibid.

23. For the use of work language analysis in HCI design, see, for example, Peter Bøgh Andersen, *A Theory of Computer Semiotics: Semiotic Approaches to Construction and Assessment of Computer Systems* (Cambridge: Cambridge University Press, 1990). For the use of metaphors as a basis for design, see John M. Carroll and John C. Thomas, "Metaphors and the Cognitive Representation of Computing Systems," *IEEE Transactions on Systems, Man, and Cybernetics* 12, no. 2 (1982): 107–116.

24. Peter Bøgh Andersen, "Ships and Movies," *Cognition, Technology, and Work* 5, no. 4 (2003): 294–301.

25. J. C. R. Licklider, "Man–Computer Symbiosis," in *The New Media Reader*, ed. Noah Wardrip-Fruin and Nick Montfort (Cambridge, MA: MIT Press, 2003), 74–82; Edmond M. Dewan, "Occipital Alpha Rhythm Eye Position and Lens Accommodation," *Nature* 214 (June 3, 1967): 975–977; Mark Weiser, "The Computer for the 21st Century," *Scientific American* (1991): 94–104.

26. Aarseth, for example, directly criticizes computer semiotics and Peter Bøgh Andersen: "When the relationship between surface sign and user is all that matters, *the unique dual materiality of the cybernetic sign process is disregarded*." Aarseth, *Cybertext*, 40.

27. Peter-Clement Woetmann, Martin Campostrini, Jonas Fritsch, Ann Luther Petersen, Søren Bro Pold, Allan Thomsen Volhøj, et al., *The Poetry Machine* (Aarhus: CAVI and Roskilde Library, 2012–), http://www.inkafterprint.dk/?page_id=45; Aarseth, *Cybertext*.

28. See http://uuuuuuuntitled.com.

29. Walter Benjamin, "Reply to Oscar A. H. Schmitz," in *Selected Writings*, ed. Michael William Jennings, Howard Eiland, and Gary Smith (Cambridge, MA: Belknap Press of Harvard University Press, 1999), 2:17.

30. Walter Benjamin, "The Author as Producer," in *Selected Writings*, ed. Michael William Jennings, Howard Eiland, and Gary Smith (Cambridge, MA: Belknap Press of Harvard University Press, 1996), 2:770.

31. Benjamin Noys, "Apocalypse, Tendency, Crisis," *Mute* 2, no. 15 (February 3, 2010), http://www.metamute.org/editorial/articles/apocalypse-tendency-crisis.

32. Benjamin, "The Author as Producer," 777.

33. Ibid., 774.

34. See, for example, the widely referenced and important essay by Walter Benjamin, "The Work of Art in the Age of Its Mechanical Reproduction," in *Selected Writings*, ed. Michael William Jennings, Marcus Bullock, Howard Eiland, and Gary Smith (Cambridge, MA: Belknap Press of Harvard University Press, 2003), 4:251–283.

35. Semantic capitalism is a term that Bruno uses to characterize the process of the capitalization of words in his *Google Adwords Happening* from 2002. Christophe Bruno, "Collective Hallucination and Capitalism 2.0: Scale-Free Elections in France," in Interface Criticism—Aesthetics beyond Buttons, ed. Christian Ulrik Andersen and Søren Pold (Aarhus: Aarhus University Press, 2011), 285. The term semio-capitalism is used by Berardi to describe financial capitalism's "virtualization of life and intelligence." Franco "Bifo" Berardi, *The Uprising: On Poetry and Finance* (Los Angeles: Semiotext[e], 2012), 14:24. See also Shoshana Zuboff, "Big Other: Surveillance Capitalism and the Prospects of an Information Civilization," *Journal of Information Technology* 30 (2015): 75–89.

36. Julian Oliver, Gordan Savičić, and Danja Vasiliev, "The Critical Engineering Manifesto," October 2011–2017, https://criticalengineering.org/.

37. Weise7 studio, http://weise7.org/.

38. *The Little Black Book of Wireless* is part of the cultural project Human Futures, and was developed with support from Aarhus University and Kunsthal Aarhus. See http://mab14.mediaarchitecture.org/exhibition/exhibits/little-black-book-wireless/.

39. Steven Johnson, *Interface Culture: How New Technology Shapes the Way We Create and Communicate* (San Francisco: Harper, 1997).

40. Furthermore, *House of Cards* was developed by the streaming platform Netflix through the use of viewer tracking, and in this way interfaces, software tracking, and quantification are also part of the production and distribution of the drama series, as pointed out by Grosser. Benjamin Grosser, *Touching Software*, 2016, http://bengrosser.com/projects/touching-software/. For more on streaming platforms, see chapter 2.

41. Philip E. Agre, "Surveillance and Capture: Two Models of Privacy," in *The New Media Reader*, ed. Noah Wardrip-Fruin and Nick Montfort (Cambridge, MA: MIT Press, 2003), 745. Wendy Chun also contends that it is this link with "habit" that creates the systems' validity, and that causality itself is formed "habitually," and based on experiences and estimations of probability, rather than logical argumentation (in this she draws on Gilles Deleuze's reading of David Hume, whom she claims is the philosopher of big data per se). Wendy Hui Kyong Chun, "How Things

Spread: Habits versus Viruses," in *Brandenburgisches Zentrum für Medienwissenschaften (ZeM)* (Potsdam, 2015).

42. Agre, "Surveillance and Capture," 743.

43. Ibid., 745–746.

44. Ibid., 748.

45. Chun uses the recommender system in Netflix as an example, and refers to the Netflix Prize (a competition for the best "collaborative filtering algorithm"). Chun, "How Things Spread."

46. This double-sidedness of the user (as a person with individual preferences and an actor in a large statistical body) is also reflected in collaborative filtering models. Yehuda Koren, a winner of the Netflix Prize, argues that accuracy in recommender systems can be achieved by combining two models of collaborative filtering: "latent factor models," which directly profile users and products, and "neighborhood models," which analyze similarities between users and products, and do not consider user-specific weights. Yehuda Koren, "Factorization Meets the Neighborhood: A Multifaceted Collaborative Filtering Model" (paper presented at the fourteenth ACM SIGKDD International Conference on Knowledge Discovery and Data Mining, Las Vegas, NV, August 24–27, 2008).

47. As Allan Blackwell, a professor of computer science at Cambridge University, expresses it, "I am more concerned with the reverse scenario. What if the human and computer cannot be distinguished because the human has become too much like a computer?" To Blackwell, the main challenge for interface design (and the HCI field) is this changed perception of the "inhumane" user. Alan F. Blackwell, "Interacting with an Inferred World: The Challenge of Machine Learning for Humane Computer Interaction" (paper presented at Critical Alternatives 2015: The Fifth Decennial Aarhus Conference, Aarhus University, Denmark, 2015).

48. Adrian Mackenzie, "The Production of Prediction: What Does Machine Learning Want?," *European Journal of Cultural Studies* 18, no. 4–5 (2015): 431.

49. Ibid., 435.

50. Ibid., 431.

51. Chun, "How Things Spread."

52. Jacques Derrida, *Of Grammatology*, corrected ed. (Baltimore: Johns Hopkins University Press, 1997).

53. Antoinette Rouvroy, "The End(s) of Critique: Data-Behaviourism vs. Due-Process," in *Privacy, Due Process, and the Computational Turn: The Philosophy of Law Meets the Philosophy of Technology*, ed. Mireille Hildebrandt and Katja de Vries (Abingdon, UK: Routledge, 2012), 151.

54. According to Chun, this is the case, for instance, in Netflix's promotion of movies that the user is not likely to like. Chun, "How Things Spread."

55. See also Søren Pold, "Interface Realisms: The Interface as Aesthetic Form," *Postmodern Culture* 15, no. 2 (2005), http://muse.jhu.edu/journals/postmodern_culture/v015/15.2pold.html.

56. Rouvroy, "The End(s) of Critique," 147–148.

57. Benjamin Grosser, *Facebook Demetricator*, 2012–, http://bengrosser.com/projects/facebook-demetricator/.

58. Benjamin Grosser, "What Do Metrics Want? How Quantification Prescribes Social Interaction on Facebook," *Computational Culture* 4 (2014), http://computationalculture.net/article/what-do-metrics-want. See also Grosser's follow-up project, *Go Rando*, which obfuscates Facebook's "reactions" (the various smileys, etc.).

59. Grosser, "What Do Metrics Want?"

60. Obviously, it is mainly when the quantification is used as a grammar of action that it must be visible, and not just a hidden structure that in more indirect ways controls the interface as well as, for example, the profiling or displaying of advertisements. As argued by Grosser, "Not all metrics collected by Facebook are visible in the interface, however. This raises the question of what specific metrics Facebook reveals to its users and which ones it keeps to itself. How do these hidden and revealed metrics differ, and what drives the company's decisions to show or conceal data? I would argue that Facebook's primary criterion for making such decisions is whether a particular metric will increase or decrease user participation." Ibid.

Chapter 2

1. Luigi Russolo, *The Art of Noise (Futurist Manifesto, 1913)* (1913; repr., New York: ubuclassics, 2004, http://www.ubu.com/historical/gb/russolo_noise.pdf; R. Murray Schafer, *The Tuning of the World* (New York: Knopf, 1977).

2. Russolo, *The Art of Noise*, 12.

3. Warren Sack, "Memory," in *Software Studies: A Lexicon*, ed. Matthew Fuller (Cambridge, MA: MIT Press, 2008), 190.

4. See Wendy Hui Kyong Chun, *Programmed Visions: Software and Memory* (Cambridge, MA: MIT Press, 2011).

5. Ben Shneiderman, "Direct Manipulation: A Step beyond Programming Languages," *Computer* 16, no. 8 (1983): 61–62.

6. Matthew Fuller, *Behind the Blip: Essays on the Culture of Software* (New York: Autonomedia, 2003).

7. Olav W. Bertelsen and Søren Pold, "Criticism as an Approach to Interface Aesthetics," in *Proceedings of the Third Nordic Conference on Human–Computer interaction* (Tampere, Finland: ACM Press, 2004), 30.

8. Geoff Cox, *Antithesis: The Dialectics of Software Art* (Aarhus: Digital Aesthetics Research Center, 2010), 135.

9. If HCI was guilty of largely overlooking how computing has developed as a cultural phenomenon, as argued above, electronic literature has similarly partly overlooked the ordinary uses of word processing. Kirschenbaum contends that "scholarly interest in the history of electronic literature has … gravitated overwhelmingly towards those authors who sought to reimagine our definitions of the literary" and "viewed the computer as a labor-making device rather than as the far more commonplace labor-saving device it was for most users." His book is therefore partly devoted to how word processing software was used by more traditional writers versus the artistic avant-garde of electronic literature. Matthew G. Kirschenbaum, *Track Changes: A Literary History of Word Processing* (Cambridge, MA: Belknap Press of Harvard University Press, 2016), 24, 25.

10. Fuller, *Behind the Blip*, 163.

11. Marco Deseriis, "<Nettime> Matthew Fuller on ATM," 2000, http://nettime.org/Lists-Archives/nettime-l-0009/msg00198.html.

12. Jörg Piringer, *nam shub*, 2006, http://joerg.piringer.net/index.php?href=namshub/namshub.xml&mtitle=software.

13. Lev Manovich has argued that office software interfaces apply avant-garde techniques such as cutting and pasting. Hence, he has also demonstrated how the office interface is aesthetic and cultural as much as it is functional. With the current cultural turn in computing, it is almost the other way around: software techniques are applied to the avant-garde, and software is aesthetic and cultural by default. Lev Manovich, "Avant-Garde as Software," http://manovich.net/index.php/projects/avant-garde-as-software.

14. Chris Anderson, "The Long Tail," *Wired*, October 10, 2004, http://www.wired.com/wired/archive/12.10/tail.html.

15. Michael Masnick, Michael Ho, Joyce Hung, and Leigh Beadon, "The Sky Is Rising!," October 2014, https://www.ccianet.org/wp-content/uploads/2014/10/Sky-Is-Rising-2014.pdf, 190.

16. On the ideas behind and development of runme.org, see Olga Goriunova, *Art Platforms and Cultural Production on the Internet* (London: Routledge, 2011). Catalogs and books were produced from each Readme festival. See, for example, Olga Goriunova and Alexei Shulgin, *Read_Me Software Art, and Cultures* (Aarhus: Digital Aesthetics Research Centre, 2004); Olga Goriunova, *Readme 100 Temporary Software Art Factory* (Dortmund: Books on Demand GmbH, 2005).

17. Runme.org, "About," 2003–, http://runme.org/about.tt2.

18. Apple, "App Store Review Guidelines," https://developer.apple.com/app-store/review/guidelines/.

19. Ibid.

20. Apple, "File System Programming Guide," https://developer.apple.com/library/mac/documentation/FileManagement/Conceptual/FileSystemProgrammingGuide/.

21. For a comprehensive overview, see App Business Blog, "Apple Store Rejection Reasons," http://iphoneincubator.com/blog/app-store/rejections.

22. Apple, "App Store Review Guidelines."

23. Molleindustria, *Phone Story*, 2011, http://phonestory.org/.

24. It is still available on the Android platform, and on Molleindustria's website, as an Adobe Flash version, though it does not work on iOS platforms.

25. Or as expressed by Cory Doctorow, the war against copyright has turned into a war against general-purpose computing. Cory Doctorow, "Lockdown: The Coming War on General-Purpose Computing," http://boingboing.net/2012/01/10/lockdown.html.

26. Henri Lefebvre, *Everyday Life in the Modern World*, trans. Sacha Rabinovitch (New York: Harper and Row, 1971), 102.

27. Gilles Deleuze, *Foucault*, trans. Seán Hand (Minneapolis: University of Minnesota Press, 1988), 23.

28. Ted Striphas, *The Late Age of Print: Everyday Book Culture from Consumerism to Control* (New York: Columbia University Press, 2011), 180–182.

29. Doctorow, "Lockdown."

30. Shoshana Zuboff, "Big Other: Surveillance Capitalism and the Prospects of an Information Civilization," *Journal of Information Technology* 30 (2015): 75.

31. Ron Hirson, "Uber: The Big Data Company," *Forbes*, March 23, 2015, http://www.forbes.com/sites/ronhirson/2015/03/23/uber-the-big-data-company/.

32. For a critique of the sharing economy and suggestion of platform cooperativism as a counter-strategy, see Trebor Scholz, *Platform Cooperativism: Challenging the Corporate Sharing Economy* (New York: Rosa Luxemburg Stiftung, 2016).

33. Indeed, the very notion of sharing and exchange as a moral requirement, which runs deeps in Western culture and is now reiterated in the sharing economy, seems to have its limits. As Wolfgang Sützl suggests, sharing "cannot be grasped in terms of rationality or exchange," and one should rather perceive sharing as irrational sacrifice and expenditure (in the spirit of Georges Bataille). Wolfgang Sützl, "On Sharing," in *Online Catalog Text for the Exhibition "Collective Making,"* ed. Kunsthal Aarhus (Aarhus: Kunsthal Aarhus, 2015).

34. The original Adbusters piece is currently not available, but is quoted in this (critical) article on the relations between Adbusters and the Occupy movement: Kirk MacDonald, "Adbusters and Occupy Wall Street: What Does the Left Really Want?," *Foundation Watch*, January 5, 2012, https://capitalresearch.org/2012/01/adbusters-and-occupy-wall-street-what-does-the-left-really-want/.

35. See, for example, Cindy Cohn, "2010: E-Book Buyer's Guide to E-Book Privacy," Electronic Frontier Foundation, December 6, 2010, https://www.eff.org/deeplinks/2010/12/2010-e-book

-buyers-guide-e-book-privacy; Free Software Foundation, "Defective by Design.Org," http://www.defectivebydesign.org.

36. See also Ted Striphas, "The Abuses of Literacy: Amazon Kindle and the Right to Read," *Communication and Critical/Cultural Studies* 7, no. 3 (2010): 297–317.

37. Walter J. Ong, *Orality and Literacy* (London: Routledge, 2012), 116.

38. Ubermorgen.com later parted with Gross and Bauch, who released their own version of the project as *Kindle'voke Ghost Writers* (http://traumawien.at/ghostwriters/).

39. Nrlnick Kencals, *You Funny Get Car*, ed. Ubermorgen.com (:(){ :|:& };:, 2012). For the original YouTube video of Justin Bieber (*Bieber after the Dentist*), see http://www.youtube.com/watch?v=upaJHS8mfP8.

40. For the origin of Bieber's parody (*David after Dentist*), see http://www.youtube.com/watch?v=txqiwrbYGrs (which has currently received more than a hundred million views).

41. Benjamin and Lukács were obviously mutually aware of and inspired by each other's work, and Lukács's reading of Balzac seems almost a demonstration of what Benjamin calls for in "The Author as Producer" (as discussed in the introduction). Georg Lukács, "II Verlorene Illusionen," in *Georg Lukács Werke 6—Probleme Des Realismus III—Der Historische Roman* (Neuwied und Berlin: Luchterhand, 1965), 474–489.

42. For a more elaborate discussion of Balzac's *Lost Illusions*, see Christian Ulrik Andersen and Søren Bro Pold, "The Lost Illusions of an Amazonian Forkbomb: What Lies beyond the Print Capitalism of the Gutenberg Galaxy?," in *Disrupting Business: Art and Activism in Times of Financial Crisis*, ed. Tatiana Bazzichelli and Geoff Cox (Brooklyn: Autonomedia, 2013), 125–142; Lukács, "II Verlorene Illusionen."

43. Manuel Portela, *Scripting Reading Motions: The Codex and the Computer as Self-Reflexive Machines* (Cambridge, MA: MIT Press, 2013), 208.

44. John Cayley and Daniel C. Howe, *The Readers Project*, http://thereadersproject.org/.

45. For a detailed explanation of the readers, see ibid.; Daniel C. Howe and John Cayley, "The Readers Project: Procedural Agents and Literary Vectors," *Leonardo* 44, no. 4 (2011): 317–324.

46. Portela, *Scripting Reading Motions*, 18. Francisco Ricardo also discusses the different readers and the way they make the method of reading strange by demonstrating it through programmed agents. Francisco J. Ricardo, *The Engagement Aesthetic: Experiencing New Media Art through Critique* (London: Bloomsbury Academic, 2013), 177–178.

47. Daniel C. Howe and John Cayley, "Reading, Writing, and Resisting: Literary Appropriation in the Readers Project," in *Proceedings of the Nineteenth International Symposium on Electric Art, ISEA2013, Sydney*, ed. Kathy Cleland, Laura Fisher, and Ross Harley (Sydney: ISEA International, 2013). Obvious examples besides the Google agents are Amazon's Whispernet (mentioned above), which tracks and collects reading behavior, and the Kindle, which has a feature where it shows other people's highlights and thereby tempts the reader to follow other people's reading.

For an art project exploring this, see Silvio Lorusso, Sebastian Schmieg, and Amazon Kindle Users, *Networked Optimization*, 2013, http://silviolorusso.com/work/networked-optimization/.

48. Portela, *Scripting Reading Motions*, 346.

49. John Cayley and Daniel C. Howe, *How It Is in Common Tongues* (Providence, RI: Natural Language Liberation Front, 2012).

50. This exhibition took place at InSpace Gallery, University of Edinburgh, November 1–17, 2012, and was associated with the Remediating the Social, ELMCIP conference. See also John Cayley, "Beginning with 'the Image' in *How It Is* When Translating Certain Processes of Digital Language Art," *Electronic Book Review* (March 1, 2015), http://www.electronicbookreview.com/thread/electropoetics/howitis.

51. Cayley and Howe, *How It Is in Common Tongues*, 4.

52. Roland Barthes, "From Work to Text," in *The Rustle of Language*, trans. Richard Howard (Berkeley: University of California Press, 1989), 56–64.

53. Ferdinand de Saussure, *Course in General Linguistics*, trans. Wade Baskin, Perry Meisel, and Haun Saussy (New York: Columbia University Press, 2011).

54. Zuboff, "Big Other," 79.

55. See also Aylin Caliskan-Islam, Joanna J. Bryson, and Arvind Narayanan, "Semantics Derived Automatically from Language Corpora Contain Human-Like Biases," *Science* 356, no. 6334 (April 14, 2017): 183–186.

56. Vannevar Bush, "As We May Think," in *The New Media Reader*, ed. Noah Wardrip-Fruin and Nick Montfort (1945; repr. Cambridge, MA: MIT Press, 2003), 44.

57. John Cayley, "Terms of Reference and Vectoralist Transgressions: Situating Certain Literary Transactions over Networked Services," *Amodern* 2 (October 2013), http://amodern.net/article/terms-of-reference-vectoralist-transgressions/.

58. Cayley and Howe, *How It Is in Common Tongues*, 30. This quotation includes the footnotes that in Cayley and Howe, contain URL references.

59. Ong, *Orality and Literacy*, 122.

60. Ibid., 128.

61. Howe and Cayley, "Reading, Writing, and Resisting."

62. John Cayley, "Pentameters: Toward the Dissolution of Certain Vectoralist Relations," *Amodern*. 2 (October 2013), http://amodern.net/article/pentameters-toward-the-dissolution-of-certain-vectoralist-relations/; Cayley, "Terms of Reference and Vectoralist Transgressions."

63. Cayley, "Terms of Reference and Vectoralist Transgressions."

64. Cayley, "Pentameters."

65. JODI, *ZYX*, 2012, 2014, http://zyx-app.com/zyx.html.

66. See the introduction to Agre's concept of grammar of action in the previous chapter.

67. JODI, *ZYX*.

68. Ibid.

69. Marcel Mauss, "Techniques of the Body," *Economy and Society* 2, no. 1 (1973): 72.

70. This quote from the rejection emphasizes that the administrator at Apple did not appreciate or understand the qualities of software art: "Specifically, we noticed the user interface of your app is not intuitive and could not be navigated." JODI, *ZYX*.

71. Erica Scourti, *The Outage* (London: Banner Repeater, 2014).

72. Erica Scourti, *Body Scan*, 2014, https://vimeo.com/111503640. *Body Scan* was exhibited at transmediale 2015 in Berlin.

73. Erica Scourti, "The Female Fool: Subversive Approaches to the Techno-Social Mediation of Femininity" (master's thesis, Central Saint Martins College of Art and Design, University of the Arts London, 2013), 9.

74. Ibid., cited from N. Katherine Hayles, "The Materiality of Informatics," *Configurations* 1, no. 1 (1993), 163.

75. Scourti, "The Female Fool," 9.

76. Gilles Deleuze, "Nineteenth Series of Humor," in *The Logic of Sense* (New York: Columbia University Press, 1990), 134–141.

77. Bernard Stiegler, "Organology of Dreams and Archi-Cinema," *Nordic Journal of Aesthetics* 24, no. 47 (2014): 10. Tertiary retentions is a development of Edmund Husserl's primary and secondary retention, meaning "the material inscription of the memory retentions in mnemotechnical mechanisms," or to put it plainly, how technology remembers. Bernard Stiegler, *Technics and Time, 3: Cinematic Time and the Question of Malaise*, Meridian (Stanford, CA: Stanford University Press, 2011), 4.

78. Max Horkheimer, Theodor W. Adorno, and Gunzelin Schmid Noerr, *Dialectic of Enlightenment: Philosophical Fragments* (Stanford, CA: Stanford University Press, 2002), 94.

79. See David M. Berry, *Critical Theory and the Digital* (London: Bloomsbury Publishing, 2014), 26; Bernard Stiegler, "Suffocated Desire, or How the Cultural Industry Destroys the Individual: Contribution to a Theory of Mass Consumption," *Parrhesia*, no. 13 (2011): 55.

80. Horkheimer, Adorno, and Schmid Noerr, *Dialectic of Enlightenment*, 113.

81. Ibid., 109.

82. Stiegler, "Suffocated Desire," 54.

83. Marshall McLuhan, *Understanding Media: The Extensions of Man* (1964; repr., London: Routledge, 1994). This is not meant as a general criticism of McLuhan, who is useful, and a pioneer of understanding media along with their relation to art and literature, but rather a criticism of his heritage, such as through the "Californian ideology" of the magazine *Wired*, which made him its "patron saint," as described by Richard Barbrook and Andy Cameron already in 1995. Richard Barbrook and Andy Cameron, "The Californian Ideology," *Mute* 1, no. 3 (September 1995), http://www.metamute.org/editorial/articles/californian-ideology.

84. Anderson, "The Long Tail."

85. Berry, *Critical Theory and the Digital*, 12.

86. Horkheimer, Adorno, and Schmid Noerr, *Dialectic of Enlightenment*, 113, 94.

87. See, for example, the following quote, which almost seems like a critical comment on Benjamin's more positive, dialectic view on montage, lacking his understanding of cinematic perception: "The montage character of the culture industry, the synthetic, controlled manner in which its products are assembled—factory-like not only in the film studio but also, virtually, in the compilation of the cheap biographies, journalistic novels, and hit songs—predisposes it to advertising: the individual moment, in being detachable, replaceable, estranged even technically from any coherence of meaning, lends itself to purposes outside the work." Ibid., 132.

88. Stiegler, *Technics and Time, 3*, 6, 37.

89. Ibid., 39. Besides Husserl, as mentioned in a note above, Stiegler's discussion of Horkheimer and Adorno includes a critical exploration of their use of Immanuel Kant's transcendental imagination and schematism, which "closed off all possibility of thinking a positive pharmacolology of the cinema—that is, of the cinematic art itself. For in fact, the cinematic *pharmakon* as art is what makes it possible to struggle against the cinema as toxic *pharmakon*." Stiegler, "Organology of Dreams and Archi-Cinema," 13. For the current discussion, however, the focus of Stiegler's critique will be kept to Horkheimer, Adorno, and Benjamin.

90. Stiegler, *Technics and Time, 3*, 39–40. Kino-Eye was the name Vertov used for both a film (1925) and his group. Dziga Vertov, "We: A Version of a Manifesto," in *The Film Factory: Russian and Soviet Cinema in Documents, 1896–1939*, ed. Richard Taylor Ian Christie (1922; repr., London: Routledge, 1994).

91. Stiegler, *Technics and Time, 3*, 229.

92. Theodor W. Adorno, Walter Benjamin, Ernst Bloch, Bertolt Brecht, and Georg Lukács. *Aesthetics and Politics* (London: Verso, 2007), 123–124.

93. Hansen describes how Benjamin's critical thinking on media and technology continues in the cultural theories of Hans Magnus Enzensberger, Oskar Negt, and Alexander Kluge. Even Adorno "resumed Benjamin's perspective, to some extent revising his earlier objections," in some of his later essays. Miriam Hansen, "Benjamin, Cinema, and Experience: 'The Blue Flower in the Land of Technology,'" *New German Critique* 40 (1987): 222, 182, 223.

94. Walter Benjamin, "The Author as Producer," in *Selected Writings*, ed. Michael William Jennings, Howard Eiland, and Gary Smith (Cambridge, MA: Belknap Press of Harvard University Press, 1996), 2:770. See the discussion in chapter 1.

95. "Die Signifikation, als einzige Leistung des Worts von Semantik zugelassen, vollendet sich im Signal. Ihr Signalcharakter verstärkt sich durch die Raschheit, mit welcher Sprachmodelle von oben her in Umlauf gesetzt werden." Max Horkheimer and Theodor W. Adorno, *Dialektik der Aufklärung—Philosophische Fragmente* (Frankfurt am Main: Fischer Taschenbuch Verlag, 2006), 174. Unfortunately the English translation translates "Signal" as "sign" and thereby loses this distinction. Horkheimer, Adorno, and Schmid Noerr, *Dialectic of Enlightenment*, 134.

96. "The layer of experience which made words human like those who spoke them has been stripped away, and in its prompt appropriation language takes on the coldness which hitherto was peculiar to billboards and the advertising sections of newspapers. Countless people use words and expressions which they either have ceased to understand at all or use only according to their behavioral functions, just as trademarks adhere all the more compulsively to their objects the less their linguistic meaning is apprehended." Horkheimer, Adorno, and Schmid Noerr, *Dialectic of Enlightenment: Philosophical Fragments*, 135.

97. Chun, *Programmed Visions*, 27, 28.

Chapter 3

1. Martijn de Waal, *The City as Interface: How New Media Are Changing the City* (Rotterdam: nai010 publishers, 2014), 20, 67.

2. Lawrence Lessig, *Free Culture: How Big Media Uses Technology and the Law to Lock Down Culture and Control Creativity* (New York: Penguin Press, 2004).

3. Jane Jacobs, *The Death and Life of Great American Cities* (New York: Vintage Books, 1961).

4. See, for example, John L. Austin's speech act theory. John L. Austin, *How to Do Things with Words: The William James Lectures Delivered at Harvard University in 1955* (Cambridge, MA: Harvard University Press, 1962).

5. Roland Barthes, "Semiology and Urbanism," in *The Semiotic Challenge* (1967; repr., Berkeley: University of California Press, 1994), 193.

6. Barthes claims that this understanding is concurrent with Jacques Derrida's perception of writing as not just a system of representing speech but also a system of signification on its own. Much later, Derrida indirectly affirms this in an interview: "All of philosophy in general, all of Western metaphysics, if one can speak in this global way of a Western metaphysics, is inscribed in architecture, which is not just the monument in stone, which gathers up in its body all the political, religious, cultural interpretations of a society." Jacques Derrida and Elisabeth Weber, *Points …: Interviews, 1974–1994* (Stanford, CA: Stanford University Press, 1995), 214.

7. Barthes, "Semiology and Urbanism," 194.

8. Ibid.

9. Ibid., 195.

10. Ibid., 199. Another relevant Oulipo text is Georges Perec's *Life, a User's Manual*, which describes ordinary life in a Parisian apartment house. The novel is read in a normal linear fashion, but its description of the apartments and inhabitants is composed through an intricate system from chess known as the knight's tour. This system gives the novel a certain nonnarrative coherence that resembles the context of an apartment house in a city—people meet accidentally and some connections occur, but without a real, traditional plot. Georges Perec and David Bellos, *Life, a User's Manual: Fictions* (London: Collins Harvill, 1988); Harry Mathews, Alastair Brotchie, and Raymond Queneau, *Oulipo Compendium* (London: Atlas Press, 1998), 170–172.

11. Writerly (scriptible) is the term Barthes uses to characterize the particular way of reading associated with the open text of modern literature. Writerly texts address the reader as an active producer of meaning rather than merely a passive consumer reconstructing the text's "proper" meaning through a reading predetermined by the author—as in the classical characterization of "readerly" texts. Roland Barthes, *S/Z* (Oxford: Blackwell, 1990), 4.

12. Roland Barthes, *The Eiffel Tower and Other Mythologies* (1979; repr., Berkeley: University of California Press, 1997). For an introductory overview on urban semiotics, see Mark Gottdiener and Alexandros Ph. Lagopoulos, *The City and the Sign: An Introduction to Urban Semiotics* (New York: Columbia University Press, 1986).

13. Barthes, "Semiology and Urbanism," 196, 195.

14. Roland Barthes, *Empire of Signs*, 1st US ed. (New York: Hill and Wang, 1982), 32.

15. As Barthes later elaborated in *Le plaisir du texte*, the text and reading is erotic. Roland Barthes, *Le plaisir du texte* (Paris: Éditions du Seuil, 1973).

16. Walter Benjamin, *One-Way Street*, in *Selected Writings*, ed. Marcus Bullock and Michael William Jennings (Cambridge, MA: Belknap Press of Harvard University Press, 1996), 1:444–488.

17. Ibid., 456.

18. Ibid.

19. See, for example, Norbert Bolz, *Am Ende der Gutenberg-Galaxis: Die neuen Kommunikationsverhältnisse* (Munich: Fink Verlag, 1993), 202ff.

20. Benjamin, *One-Way Street*, 476.

21. Barthes, "Semiology and Urbanism," 195–196.

22. Benjamin, *One-Way Street*, 456.

23. Norman M. Klein, *The Vatican to Vegas: A History of Special Effects* (New York: New Press, 2004).

24. Norman M. Klein and Madeleine Aktypi, "Scripting the Invisible: Conversations avec les vivants,"2005, http://www.le-hub.org/lang/en/archives/230.

25. See chapter 2.

26. Barthes, "Semiology and Urbanism," 194.

27. Spiro Kostof, *The City Shaped: Urban Patterns and Meanings through History* (London: Thames and Hudson, 1991), 271–272.

28. Erkki Huhtamo, *Illusions in Motion Media Archaeology of the Moving Panorama and Related Spectacles* (Cambridge, MA: MIT Press, 2013).

29. M. Christine Boyer, *The City of Collective Memory: Its Historical Imagery and Architectural Entertainments* (Cambridge, MA: MIT Press, 1994), 251–252.

30. Walter Benjamin and Rolf Tiedemann, *The Arcades Project* (Cambridge, MA: Belknap Press of Harvard University, 1999), 532.

31. Boyer, *The City of Collective Memory*, 41.

32. Martin Zerlang, "Potsdamer Platz—Myten Om Midten," *K&K—Kultur og Klasse* 38, no. 109 (2010): 48.

33. Boyer, *The City of Collective Memory*, 46ff.; Søren Pold, *Ex Libris Medierealistisk Litteratur: Paris, Los Angeles, and Cyberspace* (Odense: Syddansk Universitetsforlag, 2004), 54ff.

34. Erkki Huhtamo, "Monumental Attractions: Toward an Archaeology of Public Media Interfaces," in *Interface Criticism—Aesthetics Beyond Buttons*, ed. Christian Ulrik Andersen and Søren Pold (Aarhus: Aarhus University Press, 2011), 23.

35. Scott McQuire, "The Politics of Public Space in the Media City," special issue, *First Monday* 4 (2006), http://firstmonday.org/article/view/1544/1459; Scott McQuire, The Media City: Theory, Culture, and Society (London: Sage, 2008). See also Christian Ulrik Andersen and Søren Pold, "The Scripted Spaces of Urban Ubiquitous Computing: The Experience, Poetics, and Politics of Public Scripted Space," *Fiberculture Journal* 19 (2011), http://nineteen.fibreculturejournal.org/fcj-133-the-scripted-spaces-of-urban-ubiquitous-computing-the-experience-poetics-and-politics-of-public-scripted-space/.

36. Mirjam Struppek, "The Social Potential of Urban Screens," *Visual Communication* 5, no. 2 (June 1, 2006): 181.

37. McQuire, "The Politics of Public Space in the Media City."

38. Struppek, "The Social Potential of Urban Screens," 183, 184.

39. Superflex, *Karlskrona2*, http://superflex.net/tools/karlskrona_2/, 1999.

40. Timothy Druckrey, "Relational Architecture: The Work of Rafael Lozano-Hemmer," in *Debates and Credits: Media/Art/Public Domain*, ed. T. Goryucheva and E. Kluitenberg (Amsterdam: De Balie Center for Culture and Politics, 2003), 69.

41. Daniele Mancini, "Relational Architecture: Interview with Rafael Lozano-Hemmer," *Unpacked*, November 30, 2006, https://unpacked.wordpress.com/2006/11/30/relational-architecture-interview-with-rafael-lozano-hemmer/.

42. Kit Galloway and Sherrie Rabinowitz, "Avantpreneur (The New Practitioner)," 1989, http://www.ecafe.com/museum/avant.html.

43. Stephan Oettermann, *The Panorama: History of a Mass Medium* (New York: Zone Books, 1997), 41.

44. This media-technological grammar often coincides with a business model that contains the data and limits accessibility. For instance, services such as Google Maps, Instagram, and Foursquare collect the data generated by users in closed systems accessible only through company-controlled application programming interfaces that might change without warning. Consequently, even though Google Maps and Instagram seem open to ordinary users, other services cannot openly use their data or rely on access. As Martin Brynskov and colleagues point out, "The interplay between the physical city … and the ecology of artefacts and services that is the result of program and platform reciprocity, is something that affects how we read, write, use and change the city." Martin Brynskov, Carlos Carvajal Bermúdez Juan, Manu Fernández, Henrik Korsgaard, Ingrid Mulder, Katarzyna Piskorek, Lea Rekow, and Martijn de Waal, *Urban Interaction Design: Towards City Making* (Amsterdam: Urban IxD Booksprint, 2014), 61.

45. Manu Luksch, *Faceless*, Amour Four Filmproduktion, Ambient Information Systems, 2007, http://www.ambienttv.net/content/?q=faceless.

46. Around the millennium, it was estimated that an average London citizen was captured approximately three hundred times by a closed-circuit television camera per day. Now it is probably even more. Michael McCahill and Clive Norris, "CCTV in London," Centre for Criminology and Criminal Justice, University of Hull, June 2002, http://www.urbaneye.net/results/ue_wp6.pdf.

47. British actor Tilda Swinton is the voice-over narrator for *Faceless*. For the film's manuscript, see http://www.ambienttv.net/content/?q=downloads_resources.

48. Luksch, *Faceless*.

49. Ibid.

50. George Orwell, *1984* (1949; repr., London: Secker and Warburg, 1959).

51. For a discussion of the general confidence in surveillance despite the many holes and errors demonstrated by the production process of the film, see Manu Luksch and Mukul Patel, "Faceless: Chasing the Data Shadow," in *Goodbye Privacy—Ars Electronica 2007*, ed. Gerfried Stocker and Christine Schöpf (Ostfildern: Hatje Cantz, 2007), 72–78.

52. For the manifesto, including reference to the relevant laws, see http://www.ambienttv.net/content/?q=dpamanifesto.

53. Luksch and Patel, "Faceless," 74.

54. Ibid.

55. Personal information from conversation.

56. A further perspective on this "realist" way of seeing, which is preoccupied with the production of reality, is also found in Ingrid Hoelzl and Rémi Maire's examination of the JPEG image format, in which, as they notice, "'photographic' is no longer a category of images based on a specific mode of production (photographic recording), but a category of images (bitmap) that displays a specific aesthetic distribution (continuous tone), that is, any image … is defacto 'photographic.'" Ingrid Hoelzl and Rémi Maire, *Softimage: Towards a New Theory of the Digital Image.* (Bristol, UK: Intellect, 2015), 6.

57. Siegfried Kracauer, *The Mass Ornament*, trans. Thomas Y. Levin (Cambridge, MA: Harvard University Press, 1995), 78.

58. Ibid., 76.

59. See Søren Pold, "An Aesthetic Criticism of the Media: The Configuration of Art, Media, and Politics in Walter Benjamin's Materialistic Aesthetics," *Parallax* 12 (1999): 26–27.

60. "The Eye 1: The Ballet Boyz," http://www.ambienttv.net/content/?q=node/132.

61. Walter Benjamin, "The Work of Art in the Age of Its Mechanical Reproduction," in *Selected Writings*, ed. Michael William Jennings, Marcus Bullock, Howard Eiland, and Gary Smith (Cambridge, MA: Belknap Press of Harvard University Press, 2003), 4:266.

62. Walter Benjamin, "Little History of Photography," in *Selected Writings*, ed. Michael William Jennings, Howard Eiland, and Gary Smith (Cambridge, MA: Belknap Press of Harvard University Press, 2003), 2:518, 519.

63. Ibid., 527.

64. This is also demonstrated in the *Surveillance Chess* project by the Zurich-based !Mediengruppe Bitnik. The project tries in vain to address surveillance camera operators and get them to play chess on their CCTV by hijacking the CCTV monitor and intercepting the signal, asking the camera operators to call back—but none of them do. !Mediengruppe Bitnik, *Surveillance Chess*, 2012, http://chess.bitnik.org/about.html.

65. Henri Lefebvre, *The Production of Space* (Oxford: Blackwell, 1991).

66. Simon Sadler, *The Situationist City* (Cambridge, MA: MIT Press, 1998), 24–25.

67. Mark Shepard, Serendipitor, 2010, V2_Institute for the Unstable Media and Eyebeam Art+Technology Center, http://serendipitor.net.

68. Guy Debord, "Introduction to a Critique of Urban Geography," *Les Lèvres Nues* 6 (1955), http://www.cddc.vt.edu/sionline/presitu/geography.html; Guy Debord, "Theory of the Dérive," *Les Lèvres Nues* 9 (1956), http://www.cddc.vt.edu/sionline/si/theory.html.

69. Graham Harwood, *Perl Routines to Manipulate London*, 2002, http://yoha.co.uk/node/508. The next chapter will return to later works from Harwood and collaborators.

70. The poem begins with "I wander thro' each charter'd street, / Near where the charter'd Thames does flow." William Blake and Geoffrey Keynes, *Songs of Innocence and of Experience: Shewing the Two Contrary States of the Human Soul, 1789–1794* (London: Oxford University Press, 1970).

71. Florian Cramer, "When Writing Executes Itself," April 21, 2003, http://cramer.pleintekst.nl/all/wenn_schrift_sich_selbst_ausf%FChrt/writing_executing_itself.pdf.

72. Ibid.

73. Stephen Graham, "Combat Zones That See: Urban Warfare as US Military Technology," in *Observant States: Geopolitics and Visual Culture*, ed. Fraser MacDonald, Rachel Hughes, and Klaus Dodds (London: I. B. Tauris, 2010), 199–223.

74. See also Florian Cramer, *Words Made Flesh—Code, Culture, Imagination* (Rotterdam: Media Design Research, Piet Zwart Institute, 2005), 66–70.

75. Matthew Fuller, "Freaks of Number," in *Engineering Culture*, ed. Geoff Cox and Joasia Krysa (Brooklyn: Autonomedia, 2005), 172.

76. Cramer, "When Writing Executes Itself."

77. The process of datafication will be further discussed in chapter 4.

78. Jacobs, *The Death and Life of Great American Cities*.

79. Annmarie Chandler, "Animating the Social: Mobile Image / Kit Galloway and Sherrie Rabinowitz," in *At a Distance: Precursors to Art and Activism on the Internet*, ed. Annmarie Chandler and Norie Neumark (Cambridge, MA: MIT Press, 2005), 171.

80. Nathaniel Tkacz, "From Open Source to Open Government: A Critique of Open Politics," *Ephemera* 12, no. 4 (2012): 386–405.

81. On the development of free software and the fork into open source in 1998, Christopher Kelty writes, "The two terms resulted in two separate kinds of narratives: the first, regarding Free Software, stretched back into the 1980s, promoting software freedom and resistance to proprietary software [and] … the second, regarding Open Source, was associated with the dotcom boom and the evangelism of the libertarian, pro-business hacker Eric Raymond, who focused on the economic value and cost savings that Open Source Software represented, including the pragmatic … use of Free Software in some of the largest online start-ups." Christopher M. Kelty, *Two Bits: The Cultural Significance of Free Software* (Durham, NC: Duke University Press, 2008), 99.

82. Cristina Ampatzidou, Matthijs Bouw, Froukje van de Klundert, Michiel de Lange, and Martijn de Waal, *The Hackable City: A Research Manifesto and Design Toolkit* (Amsterdam: Amsterdam Creative Industries Publishing, Rose Leighton, 2015).

83. Adam Greenfield, *Everyware: The Dawning Age of Ubiquitous Computing* (Berkeley, CA: Peachpit Press 2006), 92.

84. Ibid.

85. Kenneth Goldsmith, *Uncreative Writing: Managing Language in the Digital Age* (New York: Columbia University Press, 2011).

86. Kelty, *Two Bits*, 3.

87. Ibid., 239.

88. This has also been discussed and applied in various degrees. Obviously free software does not guarantee the actual formation of a qualified and democratic recursive public; there is a risk that it mainly includes the geeks, and it is a constant struggle to figure out ways to develop free software in relation to the urban. See, for example, Matthew Fuller and Usman Haque, *Urban Versioning System 1.0*, ed. Omar Khan, Trebor Scholz, and Mark Shepard, vol. 2 (New York: Architectural League of New York, 2008); Adam Greenfield and Mark Shepard, *Urban Computing and Its Discontents*, ed. Omar Khan, Trebor Scholz, and Mark Shepard, vol. 1 (New York: Architectural League of New York, 2007).

89. Kelty, *Two Bits*, 310, 303.

90. In Sennett's view, this sensibility related to the actual use of the city is, for instance, often reflected in modern city planning's desire to maintain urban cultural use (i.e., the social or historical context of a urban environments), and regulate anything that "sticks out, offends, or challenges." Richard Sennett, "The Open City," *Urban Age* (November 2006): 1–5, http://downloads.lsecities.net/0_downloads/Berlin_Richard_Sennett_2006-The_Open_City.pdf.

91. Ibid., 2.

92. This opposition bears a resemblance to discussions within free and open software systems—and perhaps Sennett's article, "The Open City," must even be seen in light of open system design's success in the software industries and proliferation into other areas.

93. Ibid., 4.

94. Ibid., 3.

Chapter 4

1. Metahaven (Daniel van der Velden and Vinca Kruk), "Captives of the Cloud: Part I," *e-flux* 37 (September 2012), http://www.e-flux.com/journal/captives-of-the-cloud-part-i/. The article mentioned by Metahaven is Sharon Eisner Gillett and Mitchell Kapor, "The Self-Governing Internet: Coordination by Design," in *Working Paper Series* (Cambridge, MA: Massachusetts Institute of Technology, 1997).

2. For instance, Google Maps rely on the same database service (Cloud Bigtable) that powers Google's search engine (and many other services). Google, "Cloud Bigtable," https://cloud.google.com/bigtable/. See also Paul Dourish, "NoSQL: The Shifting Materialities of Database Technology," *Computational Culture* 4 (2014), http://computationalculture.net/article/no-sql-the-shifting-materialities-of-database-technol.

3. Metahaven (Daniel van der Velden and Vinca Kruk), "Captives of the Cloud: Part II," *e-flux* 38 (October 2012), http://www.e-flux.com/journal/captives-of-the-cloud-part-ii/.

4. Metahaven, "Captives of the Cloud: Part I."

5. Ibid., 11. Also, Benjamin Bratton has described the interface industry as a "Cloud Layer." Benjamin H. Bratton, *The Stack: On Software and Sovereignty* (Cambridge, MA: MIT Press, 2015).

6. Metahaven, "Captives of the Cloud: Part I."

7. Cory Arcangel, *Super Mario Clouds*, 2002, http://www.coryarcangel.com/things-i-made/2002-001-super-mario-clouds.

8. On art and ways of seeing, see John Berger, *Ways of Seeing* (London: Penguin, 2008).

9. Olia Lialina and Dragan Espenschied, "Do You Believe in Users?," in *Digital Folklore Reader*, ed. Olia Lialina and Dragan Espenschied (Stuttgart: merz and solitude, 2009), 11.

10. Olia Lialina, "A Vernacular Web," in *Digital Folklore Reader*, ed. Olia Lialina and Dragan Espenschied (Stuttgart: merz and solitude, 2009), 27.

11. Olia Lialina, *Summer*, 2013, http://art.teleportacia.org/olia/summer/.

12. Quoted in Jillian Steinhauer, "Sometimes a Gif Is All You Need," *Hyperallergic*, August 6, 2013, http://hyperallergic.com/77693/sometimes-a-gif-is-all-you-need/.

13. For Shulgin's call for participation, see https://nettime.org/Lists-Archives/nettime-l-9610/msg00000.html.

14. Lialina and Espenschied, "Do You Believe in Users?," 11.

15. Olia Lialina, *Agatha Appears*, 1997, http://www.c3.hu/collection/agatha/.

16. Geoff Cox, *Real-Time for Pirate Cinema* (Ljubljana: Aksioma | Institute for Contemporary Art, Ljubljana, 2015). The project exists as an installation, performance, online work, and book. Nicolas Maigret, Brendan Howell, and Jean-Marie Boyer, *The Pirate Cinema*, 2012–, http://thepiratecinema.com/.

17. Jeremy Morris, "Sounds in the Cloud: Cloud Computing and the Digital Music Commodity," *First Monday* 16, no. 5 (May 2, 2011), http://firstmonday.org/ojs/index.php/fm/article/view/3391/2917.

18. Ibid.

19. Lialina and Espenschied, "Do You Believe in Users?," 11.

20. The French philosopher of science and technology Bruno Latour calls for a critical realism that looks for the "whole machinery" in a quote that echoes Bertolt Brecht's Epic theater and Walter Benjamin's concept of tendency (discussed in chapter 1): "A matter of concern is what happens to a matter of fact when you add to it its whole scenography, much like you would do by shifting your attention from the stage to the whole machinery of a theatre." Bruno Latour, "What Is the Style of Matters of Concern" (Assen: Van Gorcum, 2008), 39. It is in this shift to the

machinery that objects become things. Latour's concept of thing is etymologically related to the Germanic and Scandinavian "tinge," meaning an assembly such as in the Norwegian Storting, Icelandic Althing, or Danish Folketing.

21. Timo Arnall, *Internet Machine*, May 13, 2014, http://www.elasticspace.com/2014/05/internet-machine.

22. Tung-Hui Hu, *A Prehistory of the Cloud* (Cambridge, MA: MIT Press, 2015), x.

23. Ibid., xxvi.

24. See chapter 1; Florian Cramer, "What Is Interface Aesthetics, or What Could It Be (Not)?," in *Interface Criticism—Aesthetics beyond Buttons*, ed. Christian Ulrik Andersen and Søren Pold (Aarhus: Aarhus University Press, 2011), 117–129.

25. Christian Ulrik Andersen and Søren Pold, eds., *Interface Criticism—Aesthetics beyond Buttons* (Aarhus: Aarhus University Press, 2011), 9.

26. For example, as seen in "software stacks," where a group of software applications work in tandem, such as Linux, Apache, MySQL, and Perl, PHP, or Python, or Windows Server, Internet Explorer, .NET, and SQL Server. See "Software Stack," Techopedia, https://www.techopedia.com/definition/27268/software-stack. See also Bratton, *The Stack*, 383–384n13.

27. Mark Weiser and others have used the words "pervasive" and "ubiquitous" to name a third wave in computing beyond mainframes and PCs. In this paradigm, computing is envisioned to calmly recede into the background of our lives. Paul Dourish and Genevieve Bell have critically compared this idea of invisible and seamless computing with the actual heterogeneous messiness of the real world of Ubicomp. Genevieve Bell and Paul Dourish, "Yesterday's Tomorrows: Notes on Ubiquitous Computing's Dominant Vision," *Personal and Ubiquitous Computing* 11, no. 2 (2007): 133–143; Paul Dourish and Genevieve Bell, *Divining a Digital Future: Mess and Mythology in Ubiquitous Computing* (Cambridge, MA: MIT Press, 2011).

28. Hu, *A Prehistory of the Cloud*, XVI. Hu compares the various layers of the Internet and cloud to a graft where "pre-existing infrastructures, such as the rail network, are like rootstock, while the newer fiber-optic cables resemble the uppermost portion." Ibid., 7.

29. Graham Harwood, Matsuko Yokokoji, and Richard Wright, *Tantalum Memorial*, 2008, http://yoha.co.uk/tantalum. *Tantalum Memorial* has been exhibited in several places including the San Jose Museum of Art (2008), Manifesta 7 Biennnial (Bolzano 2009), Ars Electronica (Linz 2009), and transmediale (Berlin 2009), where it was awarded first prize. For information about the work, see ibid. See also the discussion of Harwood's earlier work, *London.pl*, in chapter 3.

30. See, for example, Nick Fagge, "Picks, Pans, and Bare Hands: How Miners in the Heart of Africa Toil in Terrible Conditions to Extract the Rare Minerals That Power Your iPhone," *Mail Online*, 2015, http://www.dailymail.co.uk/news/article-3280872/iPhone-mineral-miners-Africa-use-bare-hands-coltan.html.

31. Telekommunisten, *thimbl*, 2010, http://telekommunisten.net/thimbl/; Telekommunisten, "Thimbl: Decentralized Microblogging," 2010, https://transmediale.de/thimbl-decentralized-microblogging.

32. Telekommunisten, *OCTO*, 2013–, http://telekommunisten.net/octo/.

33. Ibid.; Baruch Gottlieb and Dmytri Kleiner, "OCTO or How the Net Was Won," *Interactions* (March–April 2016): 33–35, http://interactions.acm.org/archive/view/march-april-2016/octo-or-how-the-net-was-won.

34. Compare the Berliner Unterwelten Association, devoted to researching and documenting Berlin's underground architectures. http://www.berliner-unterwelten.de/verein/forschungsthema-untergrund/netzwerke/die-berliner-stadtrohrpost.html.

35. Gottlieb and Kleiner, "OCTO or How the Net Was Won," 35.

36. Ibid.

37. Margaret Cohen, "Walter Benjamin's Phantasmagoria," *New German Critique* 48 (1989): 88.

38. Quoted in Peter Osborne, "The Postconceptual Condition," *Radical Philosophy* 184 (2014): 21.

39. Hu, *A Prehistory of the Cloud*, XIII.

40. See also Benjamin Grosser's work, discussed in chapter 1. Benjamin Grosser, *Facebook Demetricator*, 2012–, http://bengrosser.com/projects/facebook-demetricator/; Benjamin Grosser, "What Do Metrics Want? How Quantification Prescribes Social Interaction on Facebook," *Computational Culture* 4 (2014), http://computationalculture.net/article/what-do-metrics-want.

41. Dropbox, "Where Does Dropbox Store My Data?," https://www.dropbox.com/help/7.

42. Hu, *A Prehistory of the Cloud*, 10.

43. Ibid., 11. This has an effect, for example, when large cloud companies such as Google argue that they are not responsible for the performance of their software, since it only implements the behavior and language of the users, and cannot be located in a specific algorithm (such as in relation to complaints on racist or misogynist search results or suggestions).

44. Ibid., 10.

45. See, for example, Medina Roshan, "Bradley Manning Pleads Guilty to Misusing Classified Data in Wikileaks Case," Reuters, February 28, 2013, http://www.reuters.com/article/us-usa-wikileaks-manning-idUSBRE91R0T720130228. Manning is transgender and has changed her name from Bradley to Chelsea.

46. WikiLeaks, "The Afghan War Diary," https://wikileaks.org/afg/.

47. For a general overview, see, for example, "Afghan War Documents Leak," Wikipedia, https://en.wikipedia.org/wiki/Afghan_War_documents_leak.

48. See, for example, Natural Performance Manager, "Ngram-Fingerprint," https://www.npmjs.com/package/ngram-fingerprint.

49. Graham Harwood, Matsuko Yokokoji, and Matthew Fuller, *Endless War*, 2011, http://yoha.co.uk/node/761. *Endless War* was commissioned for the Void Gallery, Derry, Northern Ireland (2011), and has been shown at the Kunsthal Aarhus in Denmark (2013), Neuberger Museum of

Art in Purchase, New York (2014), Center for New Media Culture Riga in Latvia (2014), and various places in the United Kingdom as part of the Data as Culture exhibition.

50. Matthew Fuller and Andrew Goffey, *Evil Media* (Cambridge, MA: MIT Press, 2012), 111.

51. Harwood, Yokokoji, and Fuller, *Endless War*. Data atomization in computing refers to how files and documents are handled in a database. Information is not only stored but also divided and packaged (even across networks) for later retrieval, thereby allowing for different access methods for the same underlying data. Olaf Vethe, *System and Method for Storing and Manipulating Data in an Information Handling System* (Aliso Viejo, CA: Birdstep Technology, 1997).

52. Harwood, Yokokoji, and Fuller, *Endless War*.

53. See the discussion of Agre in chapter 1.

54. Harwood, Yokokoji, and Fuller, *Endless War*.

55. Moll's *CO2GLE* is a work that makes the concealed and displaced climatic effects of Google searches visible. See the introduction.

56. See, for example, Accenture and WSP Environment and Energy's news release of its Microsoft commissioned study: "Microsoft, Accenture, and WSP Environment and Energy Study Shows Significant Energy and Carbon Emissions Reduction Potential from Cloud Computing," Accenture, November 4, 2010, https://newsroom.accenture.com/article_display.cfm?article_id=5089.

57. For an extended elaboration of the scandal, see "Volkswagen Emissions Scandal," Wikipedia, https://en.wikipedia.org/wiki/Volkswagen_emissions_scandal.

58. *Nuage Vert* was later also shown in Saint-Ouen and Ivry-sur-Seine, France. For documentation of this and other works from HeHe, see http://hehe.org.free.fr/hehe/.

59. Alessandro Ludovico, "Hehe Interview," *Neural* (2009): 40.

60. Roderick Coover and Scott Rettberg, *Toxi•City: A Climate Change Narrative*, 2014, 2016, CRchange, http://www.crchange.net/toxicity/.*Toxi•City* has been exhibited at, for example, the Electronic Literature Organization's Hold the Light exhibition (Milwaukee, 2014), public libraries in Bergen (2014) and Aarhus (Aaby, 2015), and Museo Ars Santa Monica (Barcelona, 2016). The following interview describes the project: Roderick Coover, David Jhave Johnston, and Scott Rettberg, "The Poetics of Combinatory Cinema: David Jhave Johnston Interviews Roderick Coover and Scott Rettberg," *2014* 4, no. 1 (2014), http://www.soundeffects.dk/article/view/20329.

61. Referring to the 2016 edition of Coover and Rettberg, *Toxi•City*, kindly provided by the authors.

62. The running of *Toxi•City* is controlled by a JavaScript, which in the 2016 version explored here runs three (out of four) beginnings. The main part is a series of narratives consisting of an A narrative, then B narrative (there are nineteen of each), followed by a chorus or death story, and after every five of these series, one of these segues is played. After sixteen series, the ending is finally played. All film sequences are chosen randomly following these rules, no sequence is repeated within a session, and some sequences are left out, thereby ensuring that similar sessions

will rarely occur. After the session, which lasts approximately forty minutes, the script reloads and restarts. The characters are described in the accompanying description, and they consist of a commercial fisher, young woman, US Federal Emergency Management Agency relief worker, middle-aged woman, pig farmer, and teenage boy. Ibid.

63. Ibid.

64. For example, when the teenage boy describes the "mean fish" that survive in the waters, he observes, "Those fish bite, and there may be some kind of sign? … I fear everything is dying." Ibid.

65. Ibid.

66. Ibid.

67. Stephan Oettermann, *Das Panorama die Geschichte eines Massenmediums* (Frankfurt am Main: Syndikat, 1980), 9; Søren Pold, "Panoramic Realism: An Early and Illustrative Passage from Urban Space to Media Space in Honoré de Balzac's Parisian Novels, *Ferragus* and *Le Père Goriot*," *Nineteenth-Century French Studies* 29, no. 1–2 (2000): 47–63. See also the discussion of panorama in chapter 3.

68. Roderick Coover, "The Digital Panorama and Cinemascapes," in *Switching Codes: Thinking through Digital Technologies in the Humanities and Arts*, ed. Thomas Bartscherer and Roderick Coover (Chicago: University of Chicago Press, 2011), 201. Coover's text is written before *Toxi•City*, and reflects some of his earlier work and interests.

69. Miriam Bratu Hansen, "Benjamin's Aura," *Critical Inquiry* 34, no. 2 (2008): 338.

70. Shelley Jackson, *Snow: A Story in Progress, Weather Permitting*, https://www.instagram.com/snowshelleyjackson/; Shelley Jackson, *Snow: A Story in Progress, Weather Permitting*, 2014–, http://www.flickr.com/photos/25935290@N04/sets/72157639539497175.

71. Ibid.

72. Shelley Jackson, *Skin*, 2003–, http://ineradicablestain.com/skin.html. *Skin* is characterized as "a mortal work of art," and the full, "real" story is "not for public consumption," but a different story has been performed at the Berkeley Art Museum's Internet Art Portal and is available on *Skin*'s website. Whereas the real story is practically unavailable to anybody but the author, Jackson, the living "words" (as she calls the tattooed persons) exist in the real world, and might meet and form new expressions and phrases. *Skin* consequently (like the works of John Cayley and Daniel Howe, for example, discussed in chapter 2) can be characterized as a writerly and readerly work, literally setting living "words" free to form their own meaning.

73. As pointed out by Anne Nacher in "The Media Materiality as a 'Dance of Agency': Performing Literary Text with Substances" (paper presented at the International Conference on Digital Media and Textuality, University of Bremen, 2016), *Snow* is related to land art, including the way land art also worked with media documentation. Nacher especially mentioned Andy Goldsworthy.

74. Jackson, *Snow*, https://www.instagram.com/snowshelleyjackson/.

75. In fact, the increase in snowstorms in the Northern Hemisphere of the globe may, paradoxically, be explained by a warmer artic climate. Usually, the arctic climate is separated from southern weathers by the jet streams. These fast-moving winds in the upper hemisphere are created by extreme temperature differences between north and south, but as the northern temperatures are decreasing, they become more porous and allow for the Artic weather to move south. See, for example, "It's Cold and My Car Is Buried in Snow. Is Global Warming Really Happening?," Union of Concerned Scientists, http://www.ucsusa.org/global_warming/science_and_impacts/science/cold-snow-climate-change.html.

76. In this way, *Snow* highlights a different relation to media. As Jussi Parikka has outlined in his book *A Geology of Media*, media is not just an immaterial part of communication but rather profoundly attached to the earth. If nature is perceived through mediations (e.g., tools for measurements), these tools themselves originate in the nature they measure. Media and nature (or "medianatures") are "co-constituting spheres, where the ties are intensively connected in material nonhuman realities as much as in relations of power, economy, and work." Jussi Parikka, *A Geology of Media* (Minneapolis: University of Minnesota Press, 2015), 14.

Chapter 5

1. Ted Friedman, *Electric Dreams: Computers in American Culture* (New York: New York University Press, 2005), 102.

2. See, for example, Christian Ulrik Andersen, "*Monopoly* and the Logic of Sensation in *Spacewar!*," in *Fun and Software: Exploring Pleasure, Paradox, and Pain in Computing*, ed. Olga Goriunova (London: Bloomsbury, 2014), 197–212.

3. See, for example, Susanne Bødker, "Third-Wave HCI, 10 Years Later—Participation and Sharing," *Interactions* 22, no. 5 (August 2015): 24–31.

4. In a leaked National Security Agency presentation (from the Edward Snowden material), Steve Jobs is portrayed holding an iPhone as the new Big Brother and the paying customers as the zombies, with reference back to the 1984 Macintosh advertisement. "Photo Gallery: Spying on Smartphones," *Der Spiegel*, September 9, 2013, http://www.spiegel.de/fotostrecke/photo-gallery-spying-on-smartphones-fotostrecke-101201.html.

5. Hartmut Winkler, *Docuverse* (Regensburg: Boer, 1997), 214.

6. Ibid., 214ff.

7. Stuart Moulthroup, "You Say You Want a Revolution," in *The New Media Reader*, ed. Noah Wardrip-Fruin and Nick Montfort (1991; repr., Cambridge, MA: MIT Press, 2003), 703.

8. See also chapter 4; Tung-Hui Hu, *A Prehistory of the Cloud* (Cambridge, MA: MIT Press, 2015).

9. For a comprehensive elaboration of Walter Benjamin's notion of tendency, see chapter 1.

10. In 1970, the so-called Mansfield Amendment to military procurement authorization in the United States limited support for research in computing to projects directly related to defense. Charlie Gere, *Digital Culture* (London: Reaktion Books, 2002), 130.

11. In other words, "there is … more to object-oriented programming than language constructs. … The Underlying conceptual framework or semantics is just as important as the language. … For object-oriented programming, the initial observation of how people understand the world has to be formulated in a theory or conceptual framework." Ole Lehrman Madsen, Birger Møller, and Kristen Nygaard, *Object-Oriented Programming in the Beta Programming Language* (New York: ACM Press, 1993), 3.

12. Pelle Ehn and Morten Kyng, "Cardboard Computers: Mocking-It-Up or Hands-on the Future," in *Design at Work: Cooperative Design of Computer Systems*, ed. Joan Greenbaum and Morten Kyng (Hillsdale, NJ: Lawrence Erlbaum Associates, 1992), 169–195. Participatory design is of course a broad field that this book will not attempt to cover. For historical overviews of the field, see, for example, Ole Sejer Iversen, Anne Marie Kanstrup, and Marianne Graves Petersen, "A Visit to the 'New Utopia': Revitalizing Democracy, Emancipation, and Quality in Co-Operative Design," in *Proceedings of the Third Nordic Conference on Human–Computer Interaction* (Tampere, Finland: ACM, 2004), 171–179; Bødker, "Third-Wave HCI, 10 Years Later"; Ditte Amund Basballe, Kim Halskov, and Nicolai Brodersen Hansen, "The Early Shaping of Participatory Design at PDC," in *Proceedings of the 14th Participatory Design Conference: Short Papers, Interactive Exhibitions, Workshops—Volume 2* (Aarhus: ACM, 2016), 21–24.

13. Anthony Dunne and Fiona Raby, "Critical Design FAQ," http://www.dunneandraby.co.uk/content/bydandr/13/0.

14. As Dunne and Raby (ibid.) write, "Critical design needs to be closer to the everyday, that's where its power to disturb comes from. Too weird and it will be dismissed as art, too normal and it will be effortlessly assimilated."As such, critical design is also a way to describe former design traditions, such as the 1960s' and 1970s' (Italian) radical design that distorted scales, or used striking colors and shapes in architecture and interior design. The avant-garde, and often ironic, kitschy, or psychedelic expressions, aimed to subvert the user's conventional and modernist perception of a stylish as well as functional object.

15. There are numerous illustrations of critique being performed as design, and in addition, critical theory and analysis have been implemented in contemporary HCI. See, for example, Olav W. Bertelsen and Søren Pold, "Criticism as an Approach to Interface Aesthetics," in *Proceedings of the Third Nordic Conference on Human–Computer Interaction* (Tampere, Finland: ACM Press, 2004); Jeffrey Bardzell and Shaowen Bardzell, "Humanistic HCI," *Synthesis Lectures on Human-Centered Informatics* 8, no. 4 (2015): 1–185.

16. Participatory design thus also has parallels to free and open-source software development, and seeks to build recursive publics (as defined by Christopher Kelty and explained in chapter 3).

17. Susanne Bødker, "When Second Wave HCI Meets Third Wave Challenges," in *Proceedings of the 4th Nordic Conference on Human–Computer Interaction: Changing Roles* (Oslo: ACM, 2006), 6.

18. Pelle Ehn, *Work-Oriented Design of Computer Artifacts* (Stockholm: Arbetslivscentrum, 1988), 392, 391.

19. Ibid., 393. In this, Ehn is also inspired by the philosopher Michael Polanyi's ideas of tacit knowledge. A more recent application of a tool-oriented approach in HCI can be found in Susanne Bødker and Clemens Nylandsted Klokmose's use of Russian activity theory to conceptualize HCI: human activity is mediated by technological artifacts, and hence HCI should be seen as a human activity in a larger ecology of technological artifacts, too, they contend. Susanne Bødker and Clemens Nylandsted Klokmose, "The Human-Artifact Model: An Activity Theoretical Approach to Artifact Ecologies," *Human–Computer Interaction* 26, no. 4 (2011): 315–371.

20. Ehn, *Work-Oriented Design of Computer Artifacts*, 395.

21. Bødker, "Third-Wave HCI, 10 Years Later," 24.

22. See, for example, Donald A Norman, "Why Interfaces Don't Work," in *The Art of Human–Computer Interface Design*, ed. Brenda Laurel (Reading, MA: Addison-Wesley, 1990), 209–220; Donald A. Norman, *The Invisible Computer: Why Good Products Can Fail, the Personal Computer Is So Complex, and Information Appliances Are the Solution* (Cambridge, MA: MIT Press, 1998).

23. Winkler, *Docuverse*.

24. Gregory L. Ulmer, "Grammatology Hypermedia," *Postmodern Culture* 1, no. 2 (1991), https://muse.jhu.edu/article/27303. Nick Montford and Noah Wardruip-Fruin also note Ulmer's critique of the tool-oriented approach to computing in their discussion of Ehn and Kyng's cardboard computers. Noah Wardrip-Fruin and Nick Montfort, *The New Media Reader* (Cambridge, MA: MIT Press, 2003), 649–650.

25. Ulmer, "Grammatology Hypermedia."

26. On interface realism and criticism, see Søren Pold, "Interface Realisms: The Interface as Aesthetic Form," Postmodern Culture 15, no. 2 (2005), http://muse.jhu.edu/journals/postmodern_culture/v015/15.2pold.html.

27. Ulmer, "Grammatology Hypermedia."

28. Gertrud Koch, "Cosmos in Film: On the Concept of Space in Walter Benjamin's 'Work of Art' Essay," in *Walter Benjamin's Philosophy: Destruction and Experience*, ed. Andrew Benjamin and Peter Osborne (Routledge, 1994), 210.

29. Walter Benjamin, "The Author as Producer," in *Selected Writings*, ed. Michael William Jennings, Howard Eiland, and Gary Smith (Cambridge, MA: Belknap Press of Harvard University Press, 1996), 2:777. A common reference to the apparatus that extends its scope from the materialist engineering perspective is found in Michel Foucault's thinking. Foucault defines the apparatus as the assemblage of "discourses, institutions, architectural forms, regulatory decisions, laws, administrative measures, scientific statements, philosophical, moral and philanthropic propositions—in short, the said as much as the unsaid. Such are the elements of the apparatus. The apparatus itself is the system of relations that can be established between these elements." Michel

Foucault and Colin Gordon, "The Confession of the Flesh," in *Power/Knowledge: Selected Interviews and Other Writings, 1972/1977* (Brighton, UK: Harvester Press, 1980), 194.

30. Benjamin, "The Author as Producer," 776, 777.

31. Richard Wright, "Data Visualization," in *Software Studies: A Lexicon*, ed. Matthew Fuller (Cambridge, MA: MIT Press, 2008), 82.32. Though they do not themselves describe it in these terms, Ehn and Kyng, one can argue, have moved beyond their former tool perspective on technology and toward a more interface critical perspective. Ehn has, together with colleagues, developed his understanding of participatory design by including Bruno Latour's actor-network theory. Kyng aims to create sustaining alternatives to commercial information technology systems within telemedicine through open source. Erling Bjögvinsson, Pelle Ehn, and Per-Anders Hillgren, "Design Things and Design Thinking: Contemporary Participatory Design Challenges," *Design Issues* 28, no. 3 (2012): 101–116; Morten Kyng, "On Creating and Sustaining Alternatives: The Case of Danish Telehealth," in *Proceedings of the Fifth Decennial Aarhus Conference on Critical Alternatives* (Aarhus: Aarhus University Press, 2015), 5–16.

33. Benjamin, "The Author as Producer," 780. See also chapter 1. Benjamin adopts the Marxist term "proletarian revolution" to address the question of ownership over the means of production.

34. Ehn and Kyng, "Cardboard Computers," 192.

35. Ibid., 194. Ehn and Dan Sjögren have also referred to this as the design of "language games," with reference to Ludwig Wittgenstein, "making it possible for users to designers to put the 'tacit dimension' of their practical social and instrumental skill into play." Pelle Ehn and Dan Sjögren, "From System Description to Scripts for Actions," in *Design at Work: Cooperative Design of Computer Systems*, ed. Joan Greenbaum and Morten Kyng (Hillsdale, NJ: Lawrence Erlbaum Associates, 1992), 253.

36. Foucault and Gordon, "The Confession of the Flesh," 195.

37. See, for example, Tiziana Terranova, *Network Culture: Politics for the Information Age* (London: Pluto Press, 2004).

38. Jacques Rancière, "Literature, Politics, Aesthetics: Approaches to Democratic Disagreement (Interview with Solange Guenoun and James H. Kavanagh)," *SubStance* 92 (2000): 7, quoted in Peter Hallward, "Jacques Rancière and the Subversion of Mastery," *Paragraph* 28, no. 1 (2005): 36.

39. Anthony Dunne, *Hertzian Tales: Electronic Products, Aesthetic Experience, and Critical Design* (Cambridge, MA: MIT Press, 2005), 21.

40. Ibid., 33, 35. The poetic function is a concept from Roman Jakobson.

41. Ibid., 35 Dunne here refers to Viktor Shklovsky from Russian formalism.

42. Dunne and Raby, "Critical Design FAQ."

43. Dunne, *Hertzian Tales*, 42.

44. Matthew Fuller, "A Means of Mutation: *Notes on I/O/D 4: The Web Stalker*," in *Behind the Blip: Essays on the Culture of Software* (New York: Autonomedia, 2003), 62.

45. Quoted in in Dunne, *Hertzian Tales*, 24.

46. Ibid.

47. Fuller, "A Means of Mutation," 63.

48. David Garcia and Geert Lovink, "The ABC of Tactical Media," nettime-l, http://www.nettime.org/Lists-Archives/nettime-l-9705/msg00096.html.

49. Ibid.

50. See also chapter 3.

51. Ted Striphas, *The Late Age of Print: Everyday Book Culture from Consumerism to Control* (New York: Columbia University Press, 2011), 3, 45.

52. Peter-Clement Woetmann, Ursula Andkjær Olsen, Martin Campostrini, Jonas Fritsch, Ann Luther Petersen, Søren Bro Pold, Allan Thomsen Volhøj, et al., *The Poetry Machine* (Aarhus: CAVI and Roskilde Library, 2012–), http://www.inkafterprint.dk/?page_id=45. Since the beginning, a number of versions have been developed through a number of projects with the Roskilde and other libraries. The first text for *The Poetry Machine* was written by Woetman in 2012 and is called *Tilfældigvis er skærmen blevet blæk* (Accidentally, the screen turned to ink). *The Poetry Machine* has been exhibited at more than fifty libraries and events in Denmark, and also in the United States, Germany, Norway, and Romania.

53. See, for instance, Jessica Pressman, "The Aesthetics of Bookishness in Twenty-First-Century Literature," *Michigan Quarterly Review* 48, no. 4 (2009), http://hdl.handle.net/2027/spo.act2080.0048.402; Silvio Lorusso, "Post-Digital Publishing Archive," http://p-dpa.net/; Alessandro Ludovico, *Post-Digital Print: The Mutation of Publishing since 1894* (Eindhoven: Onomatopee, 2012); Christian Ulrik Andersen and Søren Bro Pold, "Post-Digital Books and Disruptive Literary Machines: Digital Literature beyond the Gutenberg and Google Galaxies," *Formules* 2014, no. 18 (2014): 164-183.

54. As a result of its exhibition at the 2012 Roskilde Festival, where about 1,000 poems were produced by the festival audience over the four days of the event, Woetmann chose 105 poems for the collection *105 Variationer* (105 variations), which is currently also available as an e-book on the Danish libraries online platform, ereolen.dk. Peter-Clement Woetmann, *105 Variationer* (Copenhagen: Arena, 2015).

55. Kenneth Goldsmith, *Uncreative Writing: Managing Language in the Digital Age* (New York: Columbia University Press, 2011).

56. For an example of the metaphoric, metonymic, and phatic dimensions, see the poem "Landscape December 1, 2014 1:42:51 PM–1:44:37 PM," http://www.inkafterprint.dk/?p=426: "On the bus, I wrote you a text message. / Where does ruined language want to go? / I try to draw your radiant eyes in my sentences. In every day, remnants of meaning slid along with me. The air and

the songs of the Earth. / And only thought of the lines, of the way they resembled, the way they colored and charmed and I don't know what."

57. For the users' observations, see Jonas Fritsch, Søren Bro Pold, Lasse Steenbock Vestergaard, and Melissa Lucas, "Ink: Designing for Performative Literary Interactions," *Personal and Ubiquitous Computing* 18, no. 7 (2014): 1551–1565. For a video of users using and reflecting on the installation, see http://media.cavi.dk/228/607868?player=true.

58. Manuel Portela, *Scripting Reading Motions: The Codex and the Computer as Self-Reflexive Machines* (Cambridge, MA: MIT Press, 2013), 206. The concept of the metareader is applied from the work of Philippe Bootz, and is applied to a reading of works by Rui Torres.

59. As such, the printing out of the individual poems is important in the way it allows users to share their poems without any monitoring from the platform as opposed to how social media function. Kathi Inman Berens, an electronic literature researcher, made this point in conversation.

60. Bernard Stiegler, *Technics and Time, 3: Cinematic Time and the Question of Malaise* (Stanford, CA: Stanford University Press, 2011), 37.

61. "Alphabet letterpress printing, in which each letter was cast on a separate piece of metal, or type, marked a psychological breakthrough of the first order. It embedded the word itself deeply in the manufacturing process and made it into a kind of commodity." Walter J. Ong, *Orality and Literacy* (London: Routledge, 2012), 116.

62. These two publication platforms were initiated by Geoff Cox and Christian Ulrik Andersen (as series editors of the publications), but the projects have involved more than a hundred participants, and the newspaper and workshops have matured in collaboration with shifting partners and the transmediale festival (Kristoffer Gansing and Tatiana Bazzichelli).

63. These workshops have taken place in collaboration with Plymouth University (2011), Universität der Künste, Berlin (2011), Leuphana Universität, Lüneburg (2012), Kunsthal Aarhus (2013), Hong Kong City University (2014), Liverpool John Moore University and Liverpool Biennial (2015), Constant, Brussels (2016).

64. For example, as assessed by STM, the leading global trade association for academic and professional publishers within the medical and technical sciences, "The total size of the global STM market in 2013 (including journals, books, technical information and standards, databases and tools, and medical communications and some related areas) was estimated … at $25.2 billion. … The market is predicted to grow at about 4% annually through 2017." Mark Ware and Michael Mabe, *The STM Report: An Overview of Scientific and Scholarly Journal Publishing* (The Hague: STM: International Association of Scientific, Technical, and Medical Publishers, 2015), 23.

65. Nishant Shah, "Back When the Past Had a Future: Being Precarious in a Network Society," *A Peer-Reviewed Newspaper* 1, no. 3 (2013): 3.

66. As Benjamin also envisioned was the purpose of the author.

67. Andy Goffey, "Technology, Logistics, and Logic: Rethinking the Problem of Fun in Software," in *Fun and Software: Exploring Pleasure, Parados, and Pain in Computing,* ed. Olga Goriunova (London: Bloomsbury, 2014), 22.

68. The Greek philosopher Diogenes also argued the same by using demonstrations and designations rather than observations. As an intervention in the Eleatic school's lack of understanding of movement and change (which was difficult to explain, as observations necessarily turned it into immobile instances), he simply rose and walked back and forth. His work thus reminds us that any replacement of a quest for essence with a performative act of designation and demonstration is bound to be critical. Søren Kierkegaard, "Gjentagelsen," in *Samlede Værker* (Copenhagen: Gyldendal, 1991), 5:115.

Bibliography

Aarseth, Espen J. *Cybertext: Perspectives on Ergodic Literature*. Baltimore: Johns Hopkins University Press, 1997.

Adorno, Theodor W., Walter Benjamin, Ernst Bloch, Bertolt Brecht, and Georg Lukács. *Aesthetics and Politics*. London: Verso, 2007. Translated from the German.

"Afghan War Documents Leak." Wikipedia. https://en.wikipedia.org/wiki/Afghan_War_documents_leak.

Agre, Philip E. "Surveillance and Capture: Two Models of Privacy." In *The New Media Reader*, edited by Noah Wardrip-Fruin and Nick Montfort, 737–760. Cambridge, MA: MIT Press, 2003.

Ampatzidou, Cristina, Matthijs Bouw, Froukje van de Klundert, Michiel de Lange, and Martijn de Waal. *The Hackable City: A Research Manifesto and Design Toolkit*. Amsterdam: Amsterdam Creative Industries Publishing, Rose Leighton, 2015.

Andaluz, César Escudero. Inter_Fight. 2015. https://escuderoandaluz.com/2015/06/08/inter_fight/.

Andersen, Christian Ulrik. "*Monopoly* and the Logic of Sensation in *Spacewar!*" In *Fun and Software: Exploring Pleasure, Paradox, and Pain in Computing*, edited by Olga Goriunova, 197–212. London: Bloomsbury, 2014.

Andersen, Christian Ulrik, and Søren Pold, eds. *Interface Criticism: Aesthetics beyond Buttons*. Aarhus: Aarhus University Press, 2011.

Andersen, Christian Ulrik, and Søren Bro Pold. "The Lost Illusions of an Amazonian Forkbomb: What Lies beyond the Print Capitalism of the Gutenberg Galaxy?" In *Disrupting Business: Art and Activism in Times of Financial Crisis*, edited by Tatiana Bazzichelli and Geoff Cox, 125–142. Brooklyn: Autonomedia, 2013.

Andersen, Christian Ulrik, and Søren Bro Pold. "Post-Digital Books and Disruptive Literary Machines: Digital Literature beyond the Gutenberg and Google Galaxies." *Formules* 2014, no. 18 (2014): 164–183.

Andersen, Christian Ulrik, and Søren Pold, eds. "The Scripted Spaces of Urban Ubiquitous Computing: The Experience, Poetics, and Politics of Public Scripted Space." *Fibreculture Journal* 19 (2011), http://nineteen.fibreculturejournal.org/fcj-133-the-scripted-spaces-of-urban-ubiquitous-computing-the-experience-poetics-and-politics-of-public-scripted-space/.

Andersen, Peter Bøgh. "Ships and Movies." *Cognition, Technology, and Work* 5, no. 4 (2003): 294–301.

Andersen, Peter Bøgh. *A Theory of Computer Semiotics: Semiotic Approaches to Construction and Assessment of Computer Systems*. Cambridge: Cambridge University Press, 1990.

Andersen, Peter Bøgh. "Vector Spaces as the Basic Component of Interactive Systems: Towards a Computer Semiotics." In *VENUS*. Aarhus: Aarhus University, 1991.

Anderson, Chris. "The Long Tail." *Wired*, October 10, 2004, http://www.wired.com/wired/archive/12.10/tail.html.

Apple. "App Store Review Guidelines." https://developer.apple.com/app-store/review/guidelines/.

Apple. "File System Programming Guide." https://developer.apple.com/library/mac/documentation/FileManagement/Conceptual/FileSystemProgrammingGuide/.

Arcangel, Cory. *Super Mario Clouds*. 2002. http://www.coryarcangel.com/things-i-made/2002-001-super-mario-clouds.

Arnall, Timo. *Internet Machine*. May 13, 2014. http://www.elasticspace.com/2014/05/internet-machine.

Austin, John L. *How to Do Things with Words: The William James Lectures Delivered at Harvard University in 1955*. Cambridge, MA: Harvard University Press, 1962.

Barbrook, Richard, and Andy Cameron. "The Californian Ideology." *Mute* 1, no. 3 (September 1995), http://www.metamute.org/editorial/articles/californian-ideology.

Bardzell, Jeffrey, and Shaowen Bardzell. "Humanistic HCI." *Synthesis Lectures on Human-Centered Informatics* 8, no. 4 (2015): 1–185.

Barthes, Roland. *The Eiffel Tower and Other Mythologies*. Berkeley: University of California Press, 1997. First published 1979.

Barthes, Roland. *Empire of Signs*. 1st US ed. New York: Hill and Wang, 1982.

Barthes, Roland. *Le Plaisir du Texte*. Paris: Éditions du Seuil, 1973.

Barthes, Roland. "Semiology and Urbanism." In *The Semiotic Challenge*, 191–201. Berkeley: University of California Press, 1994. First published 1967.

Barthes, Roland. *S/Z*. Oxford: Blackwell, 1990.

Barthes, Roland. "From Work to Text." In *The Rustle of Language*, translated by Richard Howard, 56–64. Berkeley: University of California Press, 1989.

Basballe, Ditte Amund, Kim Halskov, and Nicolai Brodersen Hansen. "The Early Shaping of Participatory Design at PDC." In *Proceedings of the 14th Participatory Design Conference: Short Papers, Interactive Exhibitions, Workshops—Volume 2*, 21–24. Aarhus: ACM, 2016.

Bell, Genevieve, and Paul Dourish. "Yesterday's Tomorrows: Notes on Ubiquitous Computing's Dominant Vision." *Personal and Ubiquitous Computing* 11, no. 2 (2007): 133–143.

Benjamin, Walter. "The Author as Producer." In *Selected Writings*, edited by Michael William Jennings, Howard Eiland, and Gary Smith, 2:768–782. Cambridge, MA: Belknap Press of Harvard University Press, 1996.

Benjamin, Walter. "Little History of Photography." In *Selected Writings*, edited by Michael William Jennings, Howard Eiland, and Gary Smith, 2:508–530. Cambridge, MA: Belknap Press of Harvard University Press, 2003.

Benjamin, Walter. *One-Way Street*. In *Selected Writings*. Edited by Marcus Bullock and Michael William Jennings, 1:444–488. Cambridge, MA: Belknap Press of Harvard University Press, 1996.

Benjamin, Walter. "Reply to Oscar A. H. Schmitz." In *Selected Writings*, edited by Michael William Jennings, Howard Eiland, and Gary Smith, 2:16–19. Cambridge, MA: Belknap Press of Harvard University Press, 1999.

Benjamin, Walter. "The Work of Art in the Age of Its Mechanical Reproduction." In *Selected Writings*, edited by Michael William Jennings, Marcus Bullock, Howard Eiland, and Gary Smith, 4:251–283. Cambridge, MA: Belknap Press of Harvard University Press, 2003.

Benjamin, Walter, and Rolf Tiedemann. *The Arcades Project*. Cambridge, MA: Belknap Press of Harvard University, 1999. Translated from the German.

Berardi, Franco "Bifo." *The Uprising: On Poetry and Finance*. Vol. 14. Los Angeles: Semiotext(e), 2012.

Berger, John. *Ways of Seeing*. London: Penguin, 2008.

Berry, David M. *Critical Theory and the Digital*. London: Bloomsbury Publishing, 2014.

Berry, David M. "The Uses of Object-Oriented Ontology." May 25, 2012. http://stunlaw.blogspot.dk/2012/05/uses-of-object-oriented-ontology.html.

Berry, David M., and Michael Dieter. *Postdigital Aesthetics: Art, Computation, and Design*. Basingstoke, UK: Palgrave Macmillan, 2015.

Bertelsen, Olav W., and Søren Pold. "Criticism as an Approach to Interface Aesthetics." In *Proceedings of the Third Nordic Conference on Human-Computer Interaction*. Tampere, Finland: ACM Press, 2004.

Bjögvinsson, Erling, Pelle Ehn, and Per-Anders Hillgren. "Design Things and Design Thinking: Contemporary Participatory Design Challenges." *Design Issues* 28, no. 3 (2012): 101–116.

Blackwell, Alan F. "Interacting with an Inferred World: The Challenge of Machine Learning for Humane Computer Interaction." Paper presented at Critical Alternatives 2015: The Fifth Decennial Aarhus Conference, Aarhus University, Denmark, 2015.

Blake, William, and Geoffrey Keynes. *Songs of Innocence and of Experience: Shewing the Two Contrary States of the Human Soul, 1789–1794*. London: Oxford University Press, 1970.

Bødker, Susanne. "Third-Wave HCI, 10 Years Later—Participation and Sharing." *Interactions* 22, no. 5 (August 2015): 24–31.

Bødker, Susanne. "When Second Wave HCI Meets Third Wave Challenges." In *Proceedings of the 4th Nordic Conference on Human-Computer Interaction: Changing Roles*. Oslo: ACM, 2006.

Bødker, Susanne, and Clemens Nylandsted Klokmose. "The Human-Artifact Model: An Activity Theoretical Approach to Artifact Ecologies." *Human–Computer Interaction* 26, no. 4 (2011): 315–371.

Bolter, Jay David, and Richard Grusin. *Remediation: Understanding New Media*. Cambridge, MA: MIT Press, 1999.

Bolz, Norbert. *Am Ende der Gutenberg—Galaxis die neuen Kommunikationsverhältnisse*. Munich: Fink Verlag, 1993.

Bower, Joseph L., and Clayton M. Christensen. "Disruptive Technologies: Catching the Wave." *Harvard Business Review* 73, no. 1 (January 1995): 43–53.

Boyer, M. Christine. *The City of Collective Memory: Its Historical Imagery and Architectural Entertainments*. Cambridge, MA: MIT Press, 1994.

Bratton, Benjamin H. *The Stack: On Software and Sovereignty*. Cambridge, MA: MIT Press, 2015.

Bruno, Christophe. Artwar(e). 2010–2012. http://www.artwar-e.biz/en/home/.

Bruno, Christophe. "Collective Hallucination and Capitalism 2.0: Scale-Free Elections in France." In *Interface Criticism—Aesthetics beyond Buttons*, edited by Christian Ulrik Andersen and Søren Pold, 279–292. Aarhus: Aarhus University Press, 2011.

Bruno, Christophe. *Fascinum*. 2001. http://www.unbehagen.com/fascinum/.

Brynskov, Martin, Carlos Carvajal Bermúdez Juan, Manu Fernández, Henrik Korsgaard, Ingrid Mulder, Katarzyna Piskorek, Lea Rekow, and Martijn de Waal. *Urban Interaction Design: Towards City Making*. Amsterdam: Urban IxD Booksprint, 2014.

Bush, Vannevar. "As We May Think." In *The New Media Reader*, edited by Noah Wardrip-Fruin and Nick Montfort, 35–47. Cambridge, MA: MIT Press, 2003. First published 1945.

Caliskan-Islam, Aylin, Joanna J. Bryson, and Arvind Narayanan. "Semantics Derived Automatically from Language Corpora Contain Human-Like Biases." *Science* 356, no. 6334 (April 14, 2017): 183–186.

Carroll, John M., and John C. Thomas. "Metaphors and the Cognitive Representation of Computing Systems." *IEEE Transactions on Systems, Man, and Cybernetics* 12, no. 2 (1982): 107–116.

Cayley, John. "Beginning with 'the Image' in *How It Is* When Translating Certain Processes of Digital Language Art." *Electronic Book Review* (March 1, 2015), http://www.electronicbookreview.com/thread/electropoetics/howitis.

Cayley, John. "Pentameters: Toward the Dissolution of Certain Vectoralist Relations." *Amodern* 2 (October 2013), http://amodern.net/article/pentameters-toward-the-dissolution-of-certain-vectoralist-relations/.

Cayley, John. "Terms of Reference and Vectoralist Transgressions: Situating Certain Literary Transactions over Networked Services." *Amodern* 2 (October 2013), http://amodern.net/article/terms-of-reference-vectoralist-transgressions/.

Cayley, John, and Daniel C. Howe. *How It Is in Common Tongues*. Providence, RI: Natural Language Liberation Front, 2012.

Cayley, John, and Daniel C. Howe. *The Readers Project*. http://thereadersproject.org/.

Chan, Chien A., André F. Gygax, Elaine Wong, Christopher A. Leckie, Ampalavanapillai Nirmalathas, and Daniel C. Kilper. "Methodologies for Assessing the Use-Phase Power Consumption and Greenhouse Gas Emissions of Telecommunications Network Services." *Environmental Science and Technology* 47, no. 1 (2013): 485–492.

Chandler, Annmarie. Animating the Social: Mobile Image / Kit Galloway and Sherrie Rabinowitz. In *At a Distance: Precursors to Art and Activism on the Internet*, edited by Annmarie Chandler and Norie Neumark, 140–151. Cambridge, MA: MIT Press, 2005.

Chun, Wendy Hui Kyong. "How Things Spread: Habits versus Viruses." In *Brandenburgisches Zentrum für Medienwissenschaften (ZeM)*. Potsdam, 2015.

Chun, Wendy Hui Kyong. *Programmed Visions: Software and Memory*. Cambridge, MA: MIT Press, 2011.

Cohen, Margaret. "Walter Benjamin's Phantasmagoria." *New German Critique* 48 (1989): 87–107.

Cohn, Cindy. "2010: E-Book Buyer's Guide to E-Book Privacy." Electronic Frontier Foundation, December 6, 2010. https://www.eff.org/deeplinks/2010/12/2010-e-book-buyers-guide-e-book-privacy.

Coover, Roderick. "The Digital Panorama and Cinemascapes." In *Switching Codes: Thinking through Digital Technologies in the Humanities and Arts*, edited by Thomas Bartscherer and Roderick Coover, 199–217. Chicago: University of Chicago Press, 2011.

Coover, Roderick, David Jhave Johnston, and Scott Rettberg. "The Poetics of Combinatory Cinema: David Jhave Johnston Interviews Roderick Coover and Scott Rettberg." *Sound Effects* 4, no. 1 (2014), http://www.soundeffects.dk/article/view/20329.

Coover, Roderick, and Scott Rettberg. *Toxi•City: A Climate Change Narrative*. CRchange, 2014, 2016. http://www.crchange.net/toxicity/.

Cox, Geoff. *Antithesis: The Dialectics of Software Art*. Aarhus: Digital Aesthetics Research Center, 2010.

Cox, Geoff. *Real-Time for Pirate Cinema*. Ljubljana: Aksioma | Institute for Contemporary Art, Ljubljana, 2015.

Cramer, Florian. *Anti-Media: Ephemera on Speculative Arts*. Rotterdam: nai010 publishers, 2013.

Cramer, Florian. "What Is Interface Aesthetics, or What Could It Be (Not)?" In *Interface Criticism—Aesthetics beyond Buttons*, edited by Christian Ulrik Andersen and Søren Pold, 117–129. Aarhus: Aarhus University Press, 2011.

Cramer, Florian. "When Writing Executes Itself." April 21, 2003. http://cramer.pleintekst.nl/all/wenn_schrift_sich_selbst_ausf%FChrt/writing_executing_itself.pdf.

Cramer, Florian. *Words Made Flesh—Code, Culture, Imagination*. Rotterdam: Media Design Research, Piet Zwart Institute, 2005.

Debord, Guy. "Introduction to a Critique of Urban Geography." *Les Lèvres Nues* 6 (1955), http://www.cddc.vt.edu/sionline/presitu/geography.html.

Debord, Guy. "Theory of the Dérive." *Les Lèvres Nues* 9 (1956), http://www.cddc.vt.edu/sionline/si/theory.html.

Debray, Régis. *Media Manifestos: On the Technological Transmission of Cultural Forms* [*Manifestes me\0301dialogiques*]. London: Verso, 1996.

Deleuze, Gilles. *Foucault*. Translated by Seán Hand. Minneapolis: University of Minnesota Press, 1988.

Deleuze, Gilles. "Nineteenth Series of Humor." In *The Logic of Sense*, 134–141. New York: Columbia University Press, 1990.

Derrida, Jacques. *Of Grammatology*. Corrected ed. Baltimore: Johns Hopkins University Press, 1997.

Derrida, Jacques, and Elisabeth Weber. *Points …: Interviews, 1974–1994. Meridian*. Stanford, CA: Stanford University Press, 1995.

de Saussure, Ferdinand. *Course in General Linguistics*. Translated by Wade Baskin, Perry Meisel and Haun Saussy. New York: Columbia University Press, 2011.

Deseriis, Marco. "<Nettime> Matthew Fuller on ATM." 2000. http://nettime.org/Lists-Archives/nettime-l-0009/msg00198.html.

de Waal, Martijn. *The City as Interface: How New Media Are Changing the City*. Rotterdam: nai010 publishers, 2014.

Dewan, Edmond M. "Occipital Alpha Rhythm Eye Position and Lens Accommodation." *Nature* 214 (June 3, 1967): 975–977.

Doctorow, Cory. "Lockdown: The Coming War on General-Purpose Computing." http://boingboing.net/2012/01/10/lockdown.html.

Dourish, Paul. "NoSQL: The Shifting Materialities of Database Technology." *Computational Culture* 4 (2014), http://computationalculture.net/article/no-sql-the-shifting-materialities-of-database-technology.

Dourish, Paul, and Genevieve Bell. *Divining a Digital Future: Mess and Mythology in Ubiquitous Computing*. Cambridge, MA: MIT Press, 2011.

Dropbox. "Where Does Dropbox Store My Data?" https://www.dropbox.com/help/7.

Druckrey, Timothy. "Relational Architecture: The Work of Rafael Lozano-Hemmer." In *Debates and Credits: Media/Art/Public Domain*, edited by T. Goryucheva and E. Kluitenberg, 69–72. Amsterdam: De Balie Center for Culture and Politics, 2003.

Dunne, Anthony. *Hertzian Tales: Electronic Products, Aesthetic Experience, and Critical Design*. Cambridge, MA: MIT Press, 2005.

Dunne, Anthony, and Fiona Raby. "Critical Design FAQ." http://www.dunneandraby.co.uk/content/bydandr/13/0.

Eggers, Dave. *The Circle: A Novel*. London: Hamish Hamilton, 2013.

Ehn, Pelle. *Work-Oriented Design of Computer Artifacts*. Stockholm: Arbetslivscentrum, 1988.

Ehn, Pelle, and Morten Kyng. Cardboard Computers: Mocking-It-Up or Hands-on the Future. In *Design at Work: Cooperative Design of Computer Systems*, edited by Joan Greenbaum and Morten Kyng, 169–195. Hillsdale, NJ: Lawrence Erlbaum Associates, 1992.

Ehn, Pelle, and Dan Sjögren. "From System Description to Scripts for Actions." In *Design at Work: Cooperative Design of Computer Systems*, edited by Joan Greenbaum and Morten Kyng, 241–268. Hillsdale, NJ: Lawrence Erlbaum Associates, 1992.

Emerson, Lori. *Reading Writing Interfaces: From the Digital to the Bookbound*. Minneapolis: University of Minnesota Press, 2014.

Fagge, Nick. "Picks, Pans, and Bare Hands: How Miners in the Heart of Africa Toil in Terrible Conditions to Extract the Rare Minerals That Power Your iPhone." *Mail Online*, 2015. http://www.dailymail.co.uk/news/article-3280872/iPhone-mineral-miners-Africa-use-bare-hands-coltan.html.

Foucault, Michel, and Colin Gordon. "The Confession of the Flesh." In *Power/Knowledge: Selected Interviews and Other Writings, 1972/1977*, 194–228. Brighton, UK: Harvester Press, 1980.

Free Software Foundation. "Defective by Design.org." http://www.defectivebydesign.org/.

Friedman, Ted. *Electric Dreams: Computers in American Culture*. New York: New York University Press, 2005.

Fritsch, Jonas, Søren Bro Pold, Lasse Steenbock Vestergaard, and Melissa Lucas. "Ink: Designing for Performative Literary Interactions." *Personal and Ubiquitous Computing* 18, no. 7 (2014): 1551–1565.

Fuller, Matthew. *Behind the Blip: Essays on the Culture of Software*. New York: Autonomedia, 2003.

Fuller, Matthew. "Freaks of Number." In *Engineering Culture*, edited by Geoff Cox and Joasia Krysa, 161–175. Brooklyn: Autonomedia, 2005.

Fuller, Matthew. "A Means of Mutation: *Notes on I/O/D 4: The Web Stalker*." In *Behind the Blip: Essays on the Culture of Software*, 51–68. New York: Autonomedia, 2003.

Fuller, Matthew, ed. *Software Studies: A Lexicon*. Cambridge, MA: MIT Press, 2008.

Fuller, Matthew, and Andrew Goffey. *Evil Media*. Cambridge, MA: MIT Press, 2012.

Fuller, Matthew, and Usman Haque. *Urban Versioning System 1.0*. Edited by Omar Khan, Trebor Scholz, and Mark Shepard. Vol. 2. New York: Architectural League of New York, 2008.

Galloway, Alexander R. *The Interface Effect*. Cambridge, UK: Polity Press, 2012.

Galloway, Kit, and Sherrie Rabinowitz. "Avantpreneur (The New Practitioner)." 1989. http://www.ecafe.com/museum/avant.html.

Garcia, David, and Geert Lovink. "The ABC of Tactical Media." nettime-l. http://www.nettime.org/Lists-Archives/nettime-l-9705/msg00096.html.

Gere, Charlie. *Digital Culture*. London: Reaktion Books, 2002.

Gillett, Sharon Eisner, and Mitchell Kapor. "The Self-Governing Internet: Coordination by Design." In *Working Paper Series*. Cambridge, MA: Massachusetts Institute of Technology, 1997.

Goffey, Andy. "Technology, Logistics, and Logic: Rethinking the Problem of Fun in Software." In *Fun and Software: Exploring Pleasure, Paradox, and Pain in Computing*, edited by Olga Goriunova, 21–40. London: Bloomsbury, 2014.

Goldsmith, Kenneth. *Uncreative Writing: Managing Language in the Digital Age*. New York: Columbia University Press, 2011.

Google. "Cloud Bigtable." https://cloud.google.com/bigtable/.

Goriunova, Olga. *Art Platforms and Cultural Production on the Internet*. London: Routledge, 2011.

Goriunova, Olga. *Readme 100 Temporary Software Art Factory. Dortmund*: Books on Demand GmbH, 2005.

Goriunova, Olga, and Alexei Shulgin. *Read_Me, Software Art, and Cultures*. Aarhus: Digital Aesthetics Research Centre, 2004.

Gottdiener, Mark, and Alexandros Ph. Lagopoulos. *The City and the Sign: An Introduction to Urban Semiotics*. New York: Columbia University Press, 1986.

Gottlieb, Baruch, and Dmytri Kleiner. "OCTO or How the Net Was Won." *Interaction* (March–April 2016): 33–35, http://interactions.acm.org/archive/view/march-april-2016/octo-or-how-the-net-was-won.

Graham, Stephen. "Combat Zones That See: Urban Warfare as US Military Technology." In *Observant States: Geopolitics and Visual Culture*, edited by Fraser MacDonald, Rachel Hughes, and Klaus Dodds, 199–223. London: I. B. Tauris, 2010.

Greenbaum, Joan, and Morten Kyng, eds. *Design at Work: Cooperative Design of Computer Systems*. Hillsdale, NJ: Lawrence Erlbaum Associates, 1992.

Greenfield, Adam. *Everyware: The Dawning Age of Ubiquitous Computing*. Berkeley, CA: Peachpit Press, 2006.

Greenfield, Adam, and Mark Shepard. *Urban Computing and Its Discontents*. Edited by Omar Khan, Trebor Scholz, and Mark Shepard. Vol. 1. New York: Architectural League of New York, 2007.

Grosser, Benjamin. *Facebook Demetricator*. 2012–. http://bengrosser.com/projects/facebook-demetricator/.

Grosser, Benjamin. Touching Software. 2016. http://bengrosser.com/projects/touching-software/.

Grosser, Benjamin. "What Do Metrics Want? How Quantification Prescribes Social Interaction on Facebook." *Computational Culture* 4 (2014), http://computationalculture.net/article/what-do-metrics-want.

Hallward, Peter. "Jacques Rancière and the Subversion of Mastery." *Paragraph* 28, no. 1 (March 2005): 26–45.

Hansen, Miriam. "Benjamin, Cinema, and Experience: 'The Blue Flower in the Land of Technology.'" *New German Critique* 40 (1987): 179–224.

Hansen, Miriam Bratu. "Benjamin's Aura." *Critical Inquiry* 34, no. 2 (2008): 336–375.

Haraway, Donna. "Situated Knowledges: The Science Question in Feminism and the Privilege of Partial Perspective." *Feminist Studies* 14, no. 3 (1988): 575–599.

Harwood, Graham. "Perl Routines to Manipulate London—London.Pl." 2002. http://yoha.co.uk/node/508, http://yoha.co.uk/node/90.

Harwood, Graham, Matsuko Yokokoji, and Matthew Fuller. *Endless War*. 2011. http://yoha.co.uk/node/761.

Harwood, Graham, Matsuko Yokokoji, and Richard Wright. *Tantalum Memorial*. 2008. http://yoha.co.uk/tantalum.

Hayles, N. Katherine. "The Materiality of Informatics." *Configurations* 1, no. 1 (1993): 147–170.

Hayles, N. Katherine. *Writing Machines*. Cambridge, MA: MIT Press, 2002.

Hirson, Ron. "Uber: The Big Data Company." *Forbes*, March 23, 2015, http://www.forbes.com/sites/ronhirson/2015/03/23/uber-the-big-data-company/.

Hoelzl, Ingrid, and Rémi Maire. *Softimage: Towards a New Theory of the Digital Image*. Bristol, UK: Intellect, 2015.

Hookway, Branden. *Interface: A Genealogy of Mediation and Control*. Cambridge, MA: MIT Press, 2014.

Horkheimer, Max, and Theodor W. Adorno. *Dialektik der Aufklärung—Philosophische Fragmente*. Frankfurt am Main: Fischer Taschenbuch Verlag, 2006.

Horkheimer, Max, Theodor W. Adorno, and Gunzelin Schmid Noerr. *Dialectic of Enlightenment: Philosophical Fragments*. Stanford, CA: Stanford University Press, 2002.

Howe, Daniel C., and John Cayley. "The Readers Project: Procedural Agents and Literary Vectors." *Leonardo* 44, no. 4 (2011): 317–324.

Howe, Daniel C., and John Cayley. Reading, Writing, and Resisting: Literary Appropriation in the Readers Project. In *Proceedings of the Nineteenth International Symposium on Electric Art, ISEA2013, Sydney*, edited by Kathy Cleland, Laura Fisher, and Ross Harley. Sydney: ISEA International, 2013.

Hu, Tung-Hui. *A Prehistory of the Cloud*. Cambridge, MA: MIT Press, 2015.

Huhtamo, Erkki. *Illusions in Motion Media Archaeology of the Moving Panorama and Related Spectacles*. Cambridge, MA: MIT Press, 2013.

Huhtamo, Erkki. "Monumental Attractions: Toward an Archaeology of Public Media Interfaces." In *Interface Criticism—Aesthetics beyond Buttons*, edited by Christian Ulrik Andersen and Søren Pold, 21–42. Aarhus: Aarhus University Press, 2011.

Huhtamo, Erkki, and Jussi Parikka, eds. *Media Archaeology: Approaches, Applications, and Implications*. Berkeley: University of California Press, 2011.

"It's Cold and My Car Is Buried in Snow. Is Global Warming Really Happening?" Union of Concerned Scientists. http://www.ucsusa.org/global_warming/science_and_impacts/science/cold-snow-climate-change.html.

Iversen, Ole Sejer, Anne Marie Kanstrup, and Marianne Graves Petersen. "A Visit to the 'New Utopia': Revitalizing Democracy, Emancipation, and Quality in Co-Operative Design." In *Proceedings of the Third Nordic Conference on Human-Computer Interaction*, 171–179. Tampere, Finland: ACM, 2004.

Jackson, Shelley. *Skin*. 2003–. http://ineradicablestain.com/skin.html.

Jackson, Shelley. *Snow: A Story* in Progress , *Weather Permitting*. 2014–. http://www.flickr.com/photos/25935290@N04/sets/72157639539497175.

Jackson, Shelley. *Snow: A Story* in Progress , *Weather Permitting*. https://www.instagram.com/snowshelleyjackson/.

Jacobs, Jane. *The Death and Life of Great American Cities*. New York: Vintage Books, 1961.

JODI. ZYX. 2012, 2014. http://zyx-app.com/zyx.html.

Johnson, Steven. *Interface Culture: How New Technology Shapes the Way We Create and Communicate*. San Francisco: Harper, 1997.

Kelty, Christopher M. *Two Bits: The Cultural Significance of Free Software*. Durham, NC: Duke University Press, 2008.

Kencals, Nrlnick. *You Funny Get Car*. Edited by Ubermorgen.com. :(){ :|:& };:, 2012.

Kierkegaard, Søren. "Gjentagelsen." In *Samlede Værker*. Vol. 5. Copenhagen: Gyldendal, 1991.

Kirschenbaum, Matthew G. *Mechanisms: New Media and the Forensic Imagination*. Cambridge, MA: MIT Press, 2008.

Kirschenbaum, Matthew G. *Track Changes: A Literary History of Word Processing*. Cambridge, MA: Belknap Press of Harvard University Press, 2016.

Klein, Norman M. *The Vatican to Vegas: A History of Special Effects*. New York: New Press, 2004.

Klein, Norman M., and Madeleine Aktypi. "Scripting the Invisible: Conversations avec les Vivants."2005. http://www.le-hub.org/lang/en/archives/230.

Koch, Gertrud. "Cosmos in Film: On the Concept of Space in Walter Benjamin's 'Work of Art' Essay." In *Walter Benjamin's Philosophy: Destruction and Experience*, edited by Andrew Benjamin and Peter Osborne. Abingdon, UK: Routledge, 1994.

Koren, Yehuda. "Factorization Meets the Neighborhood: A Multifaceted Collaborative Filtering Model." Paper presented at the fourteenth ACM SIGKDD International Conference on Knowledge Discovery and Data Mining, Las Vegas, NV, August 24–27, 2008.

Kostof, Spiro. *The City Shaped: Urban Patterns and Meanings through History*. London: Thames and Hudson, 1991.

Kracauer, Siegfried. *The Mass Ornament*. Translated by Thomas Y. Levin. Cambridge, MA: Harvard University Press, 1995.

Kyng, Morten. "On Creating and Sustaining Alternatives: The Case of Danish Telehealth." In *Proceedings of the Fifth Decennial Aarhus Conference on Critical Alternatives*, 5–16. Aarhus: Aarhus University Press, 2015.

Landow, George P. *Hypertext: The Convergence of Contemporary Critical Theory and Technology*. Baltimore: Johns Hopkins University Press, 1992.

Latour, Bruno. *What Is the Style of Matters of Concern?* Assen: Van Gorcum, 2008.

Latour, Bruno. "Why Has Critique Run out of Steam? From Matters of Fact to Matters of Concern." *Critical Inquiry* 30, no. 2 (2004): 225–248.

Latour, Bruno, and Peter Weibel. *Making Things Public.* Cambridge, MA: MIT Press, 2005.

Lefebvre, Henri. *Everyday Life in the Modern World.* Translated by Sacha Rabinovitch. New York: Harper and Row, 1971.

Lefebvre, Henri. *The Production of Space.* Oxford: Blackwell, 1991.

Lessig, Lawrence. *Free Culture: How Big Media Uses Technology and the Law to Lock Down Culture and Control Creativity.* New York: Penguin Press, 2004.

Lialina, Olia. *Agatha Appears.* 1997. http://www.c3.hu/collection/agatha/.

Lialina, Olia. *Summer.* 2013. http://art.teleportacia.org/olia/summer/.

Lialina, Olia. "A Vernacular Web." In *Digital Folklore Reader,* edited by Olia Lialina and Dragan Espenschied. 19–33. Stuttgart: merz and solitude, 2009.

Lialina, Olia, and Dragan Espenschied. "Do You Believe in Users?" In *Digital Folklore Reader,* edited by Olia Lialina and Dragan Espenschied, 9–13. Stuttgart: merz and solitude, 2009.

Licklider, J. C. R. "Man–Computer Symbiosis." In *The New Media Reader,* edited by Noah Wardrip-Fruin and Nick Montfort, 74–82. Cambridge, MA: MIT Press, 2003.

Lorusso, Silvio. "Post-Digital Publishing Archive." http://p-dpa.net/.

Lorusso, Silvio, Sebastian Schmieg, and Amazon Kindle Users. *Networked Optimization.* 2013. http://silviolorusso.com/work/networked-optimization/.

Lovink, Geert. *Zero Comments: Blogging and Critical Internet Culture.* New York: Routledge, 2008.

Ludovico, Alessandro. "Hehe Interview." *Neural* (2009): 39–40.

Ludovico, Alessandro. *Post-Digital Print: The Mutation of Publishing since 1894.* Eindhoven: Onomatopee, 2012.

Lukács, Georg. "II Verlorene Illusionen." In *Georg Lukács Werke 6—Probleme Des Realismus III—Der Historische Roman,* 474–489. Neuwied und Berlin: Luchterhand, 1965.

Luksch, Manu. *Faceless.* Amour Four Filmproduktion, Ambient Information Systems, 2007. http://www.ambienttv.net/content/?q=faceless.

Luksch, Manu, and Mukul Patel. "Faceless: Chasing the Data Shadow." In *Goodbye Privacy—Ars Electronica 2007,* edited by Gerfried Stocker and Christine Schöpf, 72–78. Ostfildern: Hatje Cantz, 2007.

MacDonald, Kirk. "Adbusters and Occupy Wall Street: What Does the Left Really Want?" *Foundation Watch,* January 5, 2012, https://capitalresearch.org/2012/01/adbusters-and-occupy-wall-street-what-does-the-left-really-want/.

Mackenzie, Adrian. "The Production of Prediction: What Does Machine Learning Want?" *European Journal of Cultural Studies* 18, no. 4–5 (2015): 429–445.

Madsen, Ole Lehrman, Birger Møller, and Kristen Nygaard. *Object-Oriented Programming in the Beta Programming Language*. New York: ACM Press, 1993.

Maigret, Nicolas, Brendan Howell, and Jean-Marie Boyer. *The Pirate Cinema*. 2012–. http://thepiratecinema.com/.

Mancini, Daniele. "Relational Architecture: Interview with Rafael Lozano-Hemmer." *Unpacked,* November 30, 2006, https://unpacked.wordpress.com/2006/11/30/relational-architecture-interview-with-rafael-lozano-hemmer/.

Manovich, Lev. "Avant-Garde as Software." http://manovich.net/index.php/projects/avant-garde-as-software.

Manovich, Lev. *The Language of New Media*. Cambridge, MA: MIT Press, 2001.

Masnick, Michael, Michael Ho, Joyce Hung, and Leigh Beadon. "The Sky Is Rising!" October 2014. https://www.ccianet.org/wp-content/uploads/2014/10/Sky-Is-Rising-2014.pdf.Mathews, Harry, Alastair Brotchie, and Raymond Queneau. *Oulipo Compendium*. London: Atlas Press, 1998.

Mauss, Marcel. "Techniques of the Body." *Economy and Society* 2, no. 1 (1973): 70–88.

McCahill, Michael, and Clive Norris. "CCTV in London." Centre for Criminology and Criminal Justice, University of Hull, June 2002. http://www.urbaneye.net/results/ue_wp6.pdf.

McLuhan, Marshall. *Understanding Media: The Extensions of Man*. London: Routledge, 1994. First published 1964.

McQuire, Scott. *The Media City: Theory, Culture, and Society*. London: Sage, 2008.

McQuire, Scott. "The Politics of Public Space in the Media City." Special issue, *First Monday* 4 (2006), http://firstmonday.org/article/view/1544/1459.

!Mediengruppe Bitnik. *Surveillance Chess*. 2012. http://chess.bitnik.org/about.html.

Metahaven (Daniel van der Velden and Vinca Kruk). "Captives of the Cloud: Part I." *e-flux* 37 (September 2012), http://www.e-flux.com/journal/captives-of-the-cloud-part-i/.

Metahaven (Daniel van der Velden and Vinca Kruk). "Captives of the Cloud: Part II." *e-flux* 38 (October 2012), http://www.e-flux.com/journal/captives-of-the-cloud-part-ii/.

"Microsoft, Accenture and WSP Environment and Energy Study Shows Significant Energy and Carbon Emissions Reduction Potential from Cloud Computing." Accenture, September 4, 2010. https://newsroom.accenture.com/article_display.cfm?article_id=5089.

Moll, Joana. *CO2GLE*. 2015. http://www.janavirgin.com/CO2/.

Molleindustria. Phone Story. 2011. http://phonestory.org/.

Morris, Jeremy. "Sounds in the Cloud: Cloud Computing and the Digital Music Commodity." *First Monday* 16, no. 5 (May 2, 2011), http://firstmonday.org/ojs/index.php/fm/article/view/3391/2917.

Moulthroup, Stuart. "You Say You Want a Revolution." In *The New Media Reader*, edited by Noah Wardrip-Fruin and Nick Montfort, 691–704. Cambridge, MA: MIT Press, 2003. First published 1991.

Nacher, Anne. "The Media Materiality as a 'Dance of Agency': Performing Literary Text with Substances." Paper presented at the International Conference on Digital Media and Textuality, University of Bremen, 2016.

Nake, Frieder. "Human–Computer Interaction: Signs and Signals Interfacing." *Languages of Design* 2 (1994): 193–205.

Natural Performance Manager. "Ngram-Fingerprint." https://www.npmjs.com/package/ngram-fingerprint.

Norman, Donald A. *The Invisible Computer: Why Good Products Can Fail, the Personal Computer Is So Complex, and Information Appliances Are the Solution*. Cambridge, MA: MIT Press, 1998.

Norman, Donald A. "Why Interfaces Don't Work." In *The Art of Human–Computer Interface Design*, edited by Brenda Laurel, 209–220. Reading, MA: Addison-Wesley, 1990.

Noys, Benjamin. "Apocalypse, Tendency, Crisis." *Mute* 2, no. 15 (February 3, 2010), http://www.metamute.org/editorial/articles/apocalypse-tendency-crisis.

Noys, Benjamin. "The Discreet Charm of Bruno Latour." In *(Mis)Readings of Marx in Continental Philosophy*, edited by Jernej Habjan and Jessica Whyte, 195–210. London: Palgrave Macmillan, 2014.

Oettermann, Stephan. *Das Panorama die Geschichte eines Massenmediums*. Frankfurt am Main: Syndikat, 1980.

Oettermann, Stephan. *The Panorama: History of a Mass Medium*. New York: Zone Books, 1997.

Oliver, Julian, Gordan Savičić, and Danja Vasiliev. "The Critical Engineering Manifesto." https://criticalengineering.org/.

Ong, Walter J. *Orality and Literacy*. London: Routledge, 2012.

Orwell, George. *1984*. London: Secker and Warburg, 1959. First published 1949.

Osborne, Peter. "The Postconceptual Condition." *Radical Philosophy* 184 (2014): 19–27.

Parikka, Jussi. *A Geology of Media*. Minneapolis: University of Minnesota Press, 2015.

Parikka, Jussi. "New Materialism as Media Theory: Medianatures and Dirty Matter." *Communication and Critical/Cultural Studies* 9, no. 1 (2012): 95–100.

Perec, Georges, and David Bellos. *Life, a User's Manual: Fictions*. London: Collins Harvill, 1988.

"Photo Gallery: Spying on Smartphones." *Der Spiegel*, September 9, 2013, http://www.spiegel.de/fotostrecke/photo-gallery-spying-on-smartphones-fotostrecke-101201.html.

Piringer, Jörg. *nam shub*. 2006. http://joerg.piringer.net/index.php?href=namshub/namshub.xml&mtitle=software.

Pold, Søren. "An Aesthetic Criticism of the Media: The Configuration of Art, Media, and Politics in Walter Benjamin's Materialistic Aesthetics." *Parallax* 12 (1999): 22–36.

Pold, Søren. *Ex Libris Medierealistisk Litteratur : Paris, Los Angeles, and Cyberspace*. Odense: Syddansk Universitetsforlag, 2004.

Pold, Søren. "Interface Realisms: The Interface as Aesthetic Form." *Postmodern Culture* 15, no. 2 (2005), http://muse.jhu.edu/journals/postmodern_culture/v015/15.2pold.html.

Pold, Søren. "Panoramic Realism: An Early and Illustrative Passage from Urban Space to Media Space in Honoré de Balzac's Parisian Novels, *Ferragus* and *Le Père Goriot*." *Nineteenth-Century French Studies* 29, no. 1–2) (2000): 47–63.

Portela, Manuel. *Scripting Reading Motions: The Codex and the Computer as Self-Reflexive Machines*. Cambridge, MA: MIT Press, 2013.

Pressman, Jessica. "The Aesthetics of Bookishness in Twenty-First-Century Literature." *Michigan Quarterly Review* 48, no. 4 (2009), http://hdl.handle.net/2027/spo.act2080.0048.402.

Rancière, Jacques. "Literature, Politics, Aesthetics: Approaches to Democratic Disagreement (Interview with Solange Guenoun and James H. Kavanagh)." *SubStance* 92 (2000): 3–24.

Ricardo, Francisco J. *The Engagement Aesthetic: Experiencing New Media Art through Critique*. London: Bloomsbury Academic, 2013.

Roshan, Medina. "Bradley Manning Pleads Guilty to Misusing Classified Data in Wikileaks Case." Reuters, February 28, 2013, http://www.reuters.com/article/us-usa-wikileaks-manning-idUSBRE91R0T720130228.

Rouvroy, Antoinette. "The End(s) of Critique: Data-Behaviourism vs. Due-Process." In *Privacy, Due Process, and the Computational Turn: The Philosophy of Law Meets the Philosophy of Technology*, edited by Mireille Hildebrandt and Katja de Vries, 143–168. Abingdon, UK: Routledge, 2012.

Runme.org. "About." 2003–. http://runme.org/about.tt2.

Russolo, Luigi. *The Art of Noise (Futurist Manifesto, 1913)*. New York: ubuclassics, 2004. First published 1913. http://www.ubu.com/historical/gb/russolo_noise.pdf.

Sack, Warren. "Memory." In *Software Studies: A Lexicon*, edited by Matthew Fuller, 184–193. Cambridge, MA: MIT Press, 2008.

Sadler, Simon. *The Situationist City*. Cambridge, MA: MIT Press, 1998.

Schafer, R. Murray. *The Tuning of the World*. New York: Knopf, 1977.

Scholz, Trebor. *Platform Cooperativism: Challenging the Corporate Sharing Economy*. New York: Rosa Luxemburg Stiftung, 2016.

Schumpeter, Joseph A. *Capitalism, Socialism, and Democracy*. New York: Harper and Brothers, 1942.

Scourti, Erica. *Body Scan*. 2014. https://vimeo.com/111503640.

Scourti, Erica. "The Female Fool: Subversive Approaches to the Techno-Social Mediation of Femininity." Master's thesis, Central Saint Martins College of Art and Design, University of the Arts London, 2013.

Scourti, Erica. *The Outage*. London: Banner Repeater, 2014.

Seldes, Gilbert. *The Seven Lively Arts*. New York: Harper and Brothers, 1924.

Sennett, Richard. "The Open City." *Urban Age* (November 2006): 1–5. http://downloads.lsecities.net/0_downloads/Berlin_Richard_Sennett_2006-The_Open_City.pdf

Shah, Nishant. "Back When the Past Had a Future: Being Precarious in a Network Society." *A Peer-Reviewed Newspaper* 1, no. 3 (January 2013): 3.

Shepard, Mark. Serendipitor. 2010. http://serendipitor.net/.

Shneiderman, Ben. "Direct Manipulation: A Step beyond Programming Languages." *Computer* 16, no. 8 (1983): 57–69.

Steinhauer, Jillian. "Sometimes a GIF Is All You Need." *Hyperallergic*, August 6, 2013, http://hyperallergic.com/77693/sometimes-a-gif-is-all-you-need/.

Stiegler, Bernard. "Organology of Dreams and Archi-Cinema." *Nordic Journal of Aesthetics* 24, no. 47 (2014): 7–37.

Stiegler, Bernard. "Suffocated Desire, or How the Cultural Industry Destroys the Individual: Contribution to a Theory of Mass Consumption." *Parrhesia*, no. 13 (2011): 52–61.

Stiegler, Bernard. *Technics and Time, 3: Cinematic Time and the Question of Malaise*. Stanford, CA: Stanford University Press, 2011.

Striphas, Ted. "The Abuses of Literacy: Amazon Kindle and the Right to Read." *Communication and Critical/Cultural Studies* 7, no. 3 (2010): 297–317.

Striphas, Ted. *The Late Age of Print: Everyday Book Culture from Consumerism to Control*. New York: Columbia University Press, 2011.

Struppek, Mirjam. "The Social Potential of Urban Screens." *Visual Communication* 5, no. 2 (June 1, 2006): 173–188.

Superflex. Karlskrona2. 1999. http://superflex.net/tools/karlskrona_2/.

Sützl, Wolfgang. "On Sharing." In online catalog text for the exhibition "Collective Making," edited by Kunsthal Aarhus. Aarhus: Kunsthal Aarhus, 2015.

Telekommunisten. OCTO. 2013–. http://telekommunisten.net/octo/.

Telekommunisten. *Thimbl.* 2010. http://telekommunisten.net/thimbl/.

Terranova, Tiziana. *Network Culture: Politics for the Information Age.* London: Pluto Press, 2004.

Tkacz, Nathaniel. "From Open Source to Open Government: A Critique of Open Politics." *Ephemera* 12, no. 4 (2012): 386–405.

Ulmer, Gregory L. "Grammatology Hypermedia." *Postmodern Culture* 1, no. 2 (1991), https://muse.jhu.edu/article/27303.

Vertov, Dziga. "We: A Version of a Manifesto." In *The Film Factory: Russian and Soviet Cinema in Documents, 1896–1939*, edited by Richard Taylor and Ian Christie. London: Routledge, 1994. First published 1922.

Vethe, Olaf. *System and Method for Storing and Manipulating Data in an Information Handling System.* Aliso Viejo, CA: Birdstep Technology, 1997.

Wardrip-Fruin, Noah, and Nick Montfort. *The New Media Reader.* Cambridge, MA: MIT Press, 2003.

Ware, Mark, and Michael Mabe. *The STM Report: An Overview of Scientific and Scholarly Journal Publishing.* The Hague: STM: International Association of Scientific, Technical, and Medical Publishers, 2015.

Weiser, Mark. "The Computer for the 21st Century." *Scientific American* (1991): 94–104.

WikiLeaks. "The Afghan War Diary." https://wikileaks.org/afg/.

Winkler, Hartmut. *Docuverse.* Regensburg: Boer, 1997.

Woetmann, Peter-Clement. *105 Variationer.* Copenhagen: Arena, 2015.

Woetmann, Peter-Clement, Ursula Andkjær Olsen, Martin Campostrini, Jonas Fritsch, Ann Luther Petersen, Søren Bro Pold, Allan Thomsen Volhøj, et al. *The Poetry Machine.* Aarhus: CAVI and Roskilde Library, 2012–. http://www.inkafterprint.dk/?page_id=45.

Wright, Richard. "Data Visualization." In *Software Studies: A Lexicon*, edited by Matthew Fuller, 78–87. Cambridge, MA: MIT Press, 2008.

Zerlang, Martin. "Potsdamer Platz—Myten Om Midten." *K&K—Kultur og Klasse* 38, no. 109 (2010): 45–54.

Zuboff, Shoshana. "Big Other: Surveillance Capitalism and the Prospects of an Information Civilization." *Journal of Information Technology* 30 (2015): 75–89.

Index

www.ingramcontent.com/pod-product-compliance
Lightning Source LLC
LaVergne TN
LVHW081258100826
845148LV00005B/913
9780262549677